Algebra: A Complete Course

Second Edition

Module E – Progress Tests

Written by Larry Collins

VideoText Interactive

Algebra: A Complete Course
Module E – Progress Tests, Second Edition

Send all inquiries to:
VideoText*Interactive*
P.O. Box 19761
Indianapolis, IN 46219

ISBN 1-887891-83-8
4 5 6 7 8 9 10 - RPInc - 040302

Table of Contents

Progress Tests

Unit VI – Second Degree Relations and Higher – Algebraic Fractions

Part A – Operations

Part B – Solving Open Sentences

Part C – Problem Solving with Algebraic Fractions

Unit VII – Relations of Rational Number Degree

Unit VIII – Quadratic Equations

Quiz Form A

Name ____________________

Class __________ Date __________ Score ______

Unit VI – Second Degree Relations and Higher – Algebraic Fractions
Part A – Operations
Lesson 1 – Simplifying

For each of the fractions in problems 1 and 2, state the domain of the fraction, noting all restricted values.

1. $\dfrac{a-3}{3a+12}$

2. $\dfrac{5x+2}{x^2-7x+12}$

Simplify each of the fractions in problems 3-10 by factoring and looking for 1's. Assume no denominator is zero.

3. $\dfrac{110}{225}$

4. $\dfrac{a}{a^6}$

5. $\dfrac{^-52y^2}{39y^2}$

6. $\dfrac{x+4}{(x+4)(x-3)}$

Name ______________________________

Unit VI, Part A, Lesson 1, Quiz Form A
– Continued –

7. $\dfrac{(5x-3)(x+9)}{(x+1)(5x-3)}$

8. $\dfrac{x^3+x^2}{2x^3-2x}$

9. $\dfrac{3m^2-9m-30}{6m-30}$

10. $\dfrac{3x^3+24x^2+48x}{9x^4-144x^2}$

Quiz Form B

Name ______________________

Class ________ Date ________ Score ______

Unit VI – Second Degree Relations and Higher – Algebraic Fractions
Part A – Operations
Lesson 1 - Simplifying

For each of the fractions in problems 1 and 2, state the domain of the fraction, noting all restricted values.

1. $\dfrac{a+5}{4a-20}$

2. $\dfrac{7x+3}{x^2-8x+15}$

Simplify each of the fractions in problems 3-10 by factoring and looking for 1's. Assume no denominator is zero.

3. $\dfrac{90}{135}$

4. $\dfrac{m^2}{m^7}$

5. $\dfrac{^-75x^3}{30x^3}$

6. $\dfrac{x-6}{(x-6)(x+3)}$

Name ______________________________

Unit VI, Part A, Lesson 1, Quiz Form B
– Continued –

7. $\dfrac{(x+7)(2x+4)}{(2x+4)(x+3)}$

8. $\dfrac{x^4 - x^3}{3x^3 - x^5}$

9. $\dfrac{2a^2 + 7a - 4}{a^2 - 16}$

10. $\dfrac{3x^2 + 18x - 21}{15x^2 - 15}$

Quiz Form A

Name ____________________

Class ________ Date ________ Score ______

Unit VI – Second Degree Relations and Higher – Algebraic Fractions
Part A – Operations
Lesson 2 – Multiplication
Lesson 3 – Division

For problems 1-6, find each product, and simplify your result, noting any restricted values.

1. $\frac{8}{15} \cdot \frac{5}{12}$

2. $\frac{x^3}{2} \cdot \frac{2}{x^5}$

3. $\frac{6m}{15n^4} \cdot \frac{30n^8}{9m^7}$

4. $\frac{y+1}{21} \cdot \frac{14}{y+1}$

5. $\frac{^-5x}{7x-14} \cdot \frac{3x-6}{15}$

6. $\frac{x^2+7x+12}{x+5} \cdot \frac{5x+25}{x+4}$

Name ____________________

Unit VI, Part A, Lessons 2 and 3, Quiz Form A
– Continued –

For problems 7-12, find each quotient, and simplify your result, noting any restricted values.

7. $\frac{3}{8} \div \frac{4}{5}$

8. $\frac{9x}{14} \div \frac{7x^2}{2}$

9. $\frac{n^2}{n+1} \div (n-1)$

10. $\frac{a+1}{a} \div \frac{{}^-3a-3}{ab}$

11. $\frac{b^2-3b}{b^6} \div \frac{b^2+b-12}{b^7}$

12. $\frac{u^2+8u+15}{12u^2} \div \frac{4u+12}{u^2+2u}$

Quiz Form B

Name ____________________

Class ________ Date ________ Score ________

Unit VI – Second Degree Relations and Higher – Algebraic Fractions
Part A – Operations
Lesson 2 – Multiplication
Lesson 3 – Division

For problems 1-6, find each product, and simplify your result, noting any restricted values.

1. $\frac{3}{14}\cdot\frac{35}{33}$

2. $\frac{a^3}{10}\cdot\frac{5}{a^2}$

3. $\frac{^-4a^5}{9b^2}\cdot\frac{12b^5}{22a^6}$

4. $\frac{12(a-2)}{a-3}\cdot\frac{a-3}{6(a+5)}$

5. $\frac{35}{6y+8}\cdot\frac{9y+12}{7}$

6. $\frac{a-7}{a-6}\cdot\frac{a^2-5a-6}{3a-21}$

Name ______________________________

Unit VI, Part A, Lessons 2 and 3, Quiz Form B
– Continued –

For problems 7-12, find each quotient, and simplify your result, noting any restricted values.

7. $\frac{7}{9} \div \frac{2}{5}$

8. $\frac{y}{3x} \div \frac{2y^2}{9}$

9. $\frac{2m}{m+2} \div (m+2)$

10. $\frac{5x+10}{x} \div \frac{x+2}{y}$

11. $\frac{a^2-2a-15}{a^7b^2} \div \frac{a^2-25}{16a^7b^3}$

12. $\frac{x^2-16}{20x^7y} \div \frac{x+4}{x^2-2x-15}$

Quiz Form A

Name ______________________

Class __________ Date __________ Score ________

Unit VI – Second Degree Relations and Higher – Algebraic Fractions
Part A – Operations
Lesson 4 – Addition and Subtraction

For problems 1-14, find all sums and differences. Then simplify your results, noting any restricted values.

1. $\frac{8a}{5}+\frac{3a}{5}$

2. $\frac{7x^3}{3}-\frac{x^3}{3}$

3. $\frac{x+3}{4}+\frac{x-1}{4}$

4. $\frac{a+9}{a}+\frac{8}{a}$

5. $\frac{3n}{2p}-\frac{n+2}{2p}$

6. $\frac{x^2}{x^2-7x+10}-\frac{4x+5}{x^2-7x+10}$

7. $\frac{5a}{8}+\frac{a}{6}$

8. $\frac{11}{4x}-\frac{3}{7x}+\frac{1}{2x}$

Name ______________________________

Unit VI, Part A, Lesson 4, Quiz Form A
– Continued –

9. $\frac{3x-1}{4}-\frac{5x+1}{6}$

10. $\frac{b+3}{8b}+\frac{4b-3}{4b}-\frac{b-3}{2b}$

11. $\frac{5}{x}-\frac{2x-1}{x+1}$

12. $\frac{3x-1}{x-4}+\frac{2x+1}{x^2-16}$

13. $\frac{{}^{-}24}{x^2-7x+10}-\frac{x+3}{5-x}$

14. $\frac{x^2+2}{x^2-5x+4}-\frac{x-2}{x-1}$

Quiz Form B

Name ______________________________

Class __________ Date ____________ Score ________

Unit VI – Second Degree Relations and Higher – Algebraic Fractions
Part A – Operations
Lesson 4 – Addition and Subtraction

For problems 1-14, find all sums and differences. Then simplify your results, noting any restricted values.

1. $\dfrac{2x^2}{7}+\dfrac{4x^2}{7}$

2. $\dfrac{b}{9}-\dfrac{10b}{9}$

3. $\dfrac{m+5}{6}+\dfrac{m-2}{6}$

4. $\dfrac{14}{15a}-\dfrac{4}{15a}$

5. $\dfrac{2y}{3}-\dfrac{y-1}{3}$

6. $\dfrac{3x}{3x^2+10x-8}-\dfrac{2}{3x^2+10x-8}$

7. $\dfrac{5x}{14}+\dfrac{x}{7}$

8. $\dfrac{7}{4x^2}+\dfrac{2}{8x}+\dfrac{1}{2x^3}$

Name ________________________________

Unit VI, Part A, Lesson 4, Quiz Form B
– Continued –

9. $\frac{b-2}{9}+\frac{2b-1}{4}$

10. $\frac{5x-1}{6}+\frac{2x-1}{4}+\frac{4-x}{3}$

11. $\frac{3}{x-2}+\frac{2}{x}$

12. $\frac{x+1}{x^2-5x+6}+\frac{x+4}{x-3}$

13. $\frac{3m+1}{m^2-25}-\frac{2m+1}{5-m}$

14. $\frac{a^2-22}{a^2+9a-22}-\frac{a-2}{a+11}$

Quiz Form A

Name ______________________

Class __________ Date __________ Score ________

Unit VI – Second Degree Relations and Higher – Algebraic Fractions
Part A – Operations
Lesson 5 – Complex Forms

Simplify each complex fraction in problems 1-4. Assume no denominator is zero.

1. $$\frac{\dfrac{a^2+4a}{9}}{\dfrac{a+4}{3}}$$

2. $$\frac{\dfrac{3x}{7}+\dfrac{1}{14}}{\dfrac{x}{2}+\dfrac{5}{7}}$$

Name ________________________________

Unit VI, Part A, Lesson 5, Quiz Form A
– Continued –

3. $$\frac{\frac{3}{x}-\frac{12}{x^2+2x}}{\frac{2}{x+2}-\frac{1}{x}}$$

4. $$\frac{\frac{2}{x}+\frac{2}{y}}{\frac{x^2-y^2}{xy}}$$

Quiz Form B

Name ______________________

Class __________ Date __________ Score ________

Unit VI – Second Degree Relations and Higher – Algebraic Fractions
Part A – Operations
Lesson 5 – Complex Forms

Simplify each complex fraction in problems 1-4. Assume no denominator is zero.

1. $$\frac{\frac{x^2+6x}{8}}{\frac{x+6}{2}}$$

2. $$\frac{5a+\frac{1}{2}}{7a+\frac{1}{2}}$$

Name ______________________________

Unit VI, Part A, Lesson 5, Quiz Form B
– Continued –

3. $$\frac{\frac{4}{a^2+4a}-\frac{3}{a}}{\frac{2}{a+4}+\frac{4}{a}}$$

4. $$\frac{\frac{3}{a}-\frac{3}{b}}{\frac{a^2-b^2}{ab}}$$

Quiz Form A

Name ____________________

Class ________ Date ________ Score ______

Unit VI – Second Degree Relations and Higher – Algebraic Fractions
Part B – Solving Open Sentences

Lesson 1 – Equations – Arithmetic Case
Lesson 2 – Equations – Algebraic Case

Solve each of the equations in problems 1 and 2.

1. $\frac{x}{4} - \frac{3x}{2} = 5$

2. $\frac{a}{24} - \frac{4}{3} = \frac{4a}{8}$

For each of the equations in problems 3 and 4, note all restricted values. Then solve each of the equations, ruling out all apparent solutions that are unacceptable.

3. $\frac{16}{x^2 - 4x} = \frac{2}{x} + \frac{x}{x-4}$

Name ______________________________

Unit VI, Part B, Lessons 1 and 2, Quiz Form A
– Continued –

4. $\dfrac{2y-1}{y^2-9y+20}=\dfrac{4}{4-y}+\dfrac{7}{y-5}$

Quiz Form B

Name ____________________

Class ________ Date ________ Score ______

Unit VI – Second Degree Relations and Higher – Algebraic Fractions
Part B – Solving Open Sentences
Lesson 1 – Equations – Arithmetic Case
Lesson 2 – Equations – Algebraic Case

Solve each of the equations in problems 1 and 2.

1. $\frac{5}{2} - \frac{y}{8} = \frac{y}{4}$

2. $\frac{2m}{5} - \frac{m}{2} = \frac{1}{10}$

For each of the equations in problems 3 and 4, note all restricted values. Then solve each of the equations, ruling out all apparent solutions that are unacceptable.

3. $\frac{x+5}{x-4} = \frac{3}{x} + \frac{36}{x^2 - 4x}$

Name ________________________________

Unit VI, Part B, Lessons 1 and 2, Quiz Form B
– Continued –

4. $\frac{13}{x^2-4}+\frac{x}{2-x}=\frac{^{-}2}{x+2}$

Quiz Form A

Name ______________________

Class __________ Date __________ Score ________

Unit VI – Second Degree Relations and Higher – Algebraic Fractions
Part B – Solving Open Sentences
Lesson 3 – Inequalities – Algebraic Case
Lesson 4 – Literal Equations

Solve each of the inequalities in problems 1 and 2 by considering the range of each factor relative to a number line. Note all restricted values, and rule out any apparent solutions that are unacceptable. Then show your solution set both on the accompanying number line and using set notation.

1. $\dfrac{5-2x}{4x+3} \le 0$

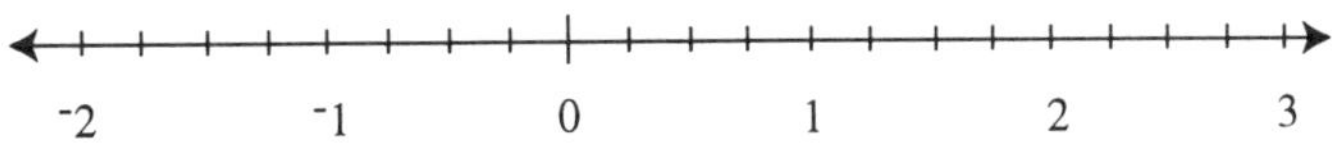

2. $\dfrac{(x+4)(x-1)}{x+3} > 0$

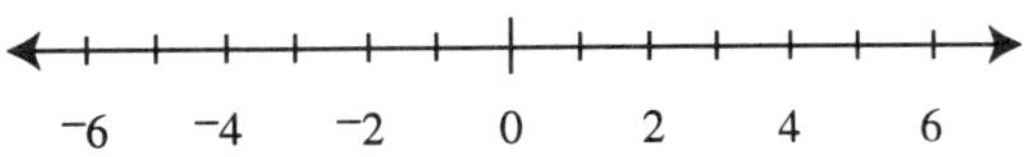

Name ___________________________________

Unit VI, Part B, Lessons 2 and 3, Quiz Form A
– Continued –

Solve each of the literal equations in problems 3 and 4 for the placeholder indicated. Assume no placeholder is zero.

3. $v = \frac{x - t}{s}$ for x

4. $\frac{rs}{a} + \frac{st}{b} = 1$ for s

Quiz Form B

Name ______________________

Class __________ Date __________ Score ______

Unit VI – Second Degree Relations and Higher – Algebraic Fractions
Part B – Solving Open Sentences
Lesson 3 – Inequalities – Algebraic Case
Lesson 4 – Literal Equations

Solve each of the inequalities in problems 1 and 2 by considering the range of each factor relative to a number line. Note all restricted values, and rule out any apparent solutions that are unacceptable. Then show your solution set both on the accompanying number line and using set notation.

1. $\dfrac{3x+2}{x-3} \le 0$

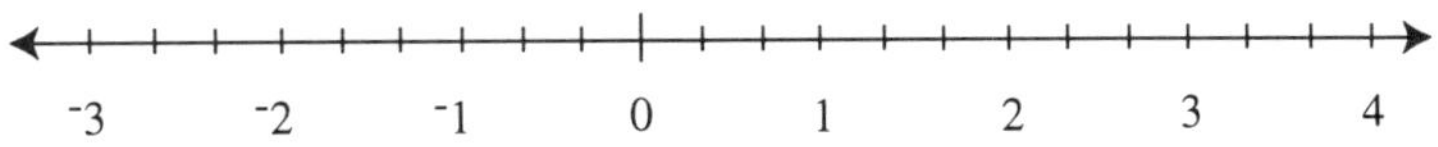

2. $\dfrac{(x-2)(x+1)}{x-5} < 0$

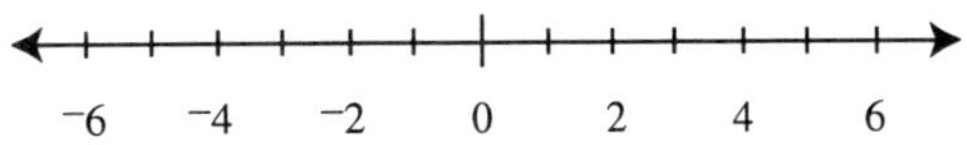

Name ______________________________

Unit VI, Part B, Lessons 2 and 3, Quiz Form B
– Continued –

Solve each of the literal equations in problems 3 and 4 for the placeholder indicated. Assume no placeholder is zero.

3. $\frac{x}{y} - g = v$ for x

4. $S = \frac{a - ar^2}{1 - r}$ for a

Quiz Form A

Name ______________________

Class ________ Date ________ Score ______

Unit VI – Second Degree Relations and Higher – Algebraic Fractions
Part C – Problem Solving with Algebraic Fractions

Lesson 1 – "Fraction" Problems
Lesson 2 – "Work" Problems
Lesson 3 – "Motion" Problems

For each of the following story problems, answer the four analysis questions to find the system of equations needed to solve. Then solve, and use common sense to check your answers.

1. Sara can mow the lawn in 3 hours. Her sister, Kim, can mow the same lawn in 2 hours. If they worked together, how long would it take to mow the lawn?

2. A number is added to the numerator and subtracted from the denominator of the fraction $\frac{2}{9}$. The resulting fraction is the reciprocal of $\frac{2}{9}$. Find the number.

Name ________________________________

Unit VI, Part C, Lessons 1, 2 and 3, Quiz Form A
– Continued –

3. Chuck drove 240 kilometers in the same amount of time that Mary drove 190 kilometers. Chuck drove 20 kilometers per hour faster than Mary. How fast did each drive?

Quiz Form B

Name ______________________

Class ________ Date ________ Score ______

Unit VI – Second Degree Relations and Higher – Algebraic Fractions
Part C – Problem Solving with Algebraic Fractions
Lesson 1 – "Fraction" Problems
Lesson 2 – "Work" Problems
Lesson 3 – "Motion" Problems

For each of the following story problems, answer the four analysis questions to find the system of equations needed to solve. Then solve, and use common sense to check your answers.

1. Jim paints a house in 6 days. It takes Sam 9 days to paint the same house. How long will it take them to paint the house if they work together?

2. The denominator of a fraction is 4 more than the numerator. If the fraction simplifies to 5, find the fraction.

Name ______________________________

Unit VI, Part C, Lessons 1, 2 and 3, Quiz Form B
– Continued –

3. Adam biked 14 kilometers in the same amount of time that Linda hiked 10 kilometers. Adam's rate was 4 kilometers per hour more than Linda's. How fast did each travel?

Quiz Form A Name ______________________

Class __________ Date __________ Score ______

Unit VI – Second Degree Relations and Higher – Algebraic Fractions
Part C – Problem Solving with Algebraic Fractions
Lesson 4 – "Direct Variation" Problems
Lesson 5 – "Inverse Variation" Problems
Lesson 6 – "Mixed Variation" Problems

For each of the following story problems about variation, use the appropriate applications of the models of variation to find the relation needed to solve. Then solve, and use common sense to check your answer.

1. The current in an electric circuit varies inversely as the resistance. When the current is 40 amps, the resistance is 25 ohms. Find the current when the resistance is 15 ohms.

2. The cost of chocolates varies directly as the number of pounds. If 2 pounds of chocolate cost \$2.80, find the cost of 7 pounds of chocolate.

Name ______________________________

Unit VI, Part C, Lessons 4, 5 and 6, Quiz Form A
– Continued –

3. The volume of a pyramid varies jointly as the height and the area of the base. A pyramid 18 cm high whose base has an area of 30 square cm has a volume of 180 cubic centimeters. What is the area of the base, if the volume increases to 200 cubic centimeters, and the height is decreased to 15 centimeters?

Quiz Form B

Name ______________________

Class __________ Date __________ Score ________

Unit VI – Second Degree Relations and Higher – Algebraic Fractions
Part C – Problem Solving with Algebraic Fractions
Lesson 4 – "Direct Variation" Problems
Lesson 5 – "Inverse Variation" Problems
Lesson 6 – "Mixed Variation" Problems

For each of the following story problems about variation, use the appropriate applications of the models of variation to find the relation needed to solve. Then solve, and use common sense to check your answer.

1. The time to travel a fixed distance varies inversely as the rate of travel. When the time traveled is 6 hours, the rate of travel is 90 kilometers per hour. Find the time when the rate of travel is 80 kilometer per hour.

2. The weight of a metal rod varies directly as its length. If a 12-foot rod weighs 18 pounds, how much does a 20-foot rod weigh?

Name ______________________________

Unit VI, Part C, Lessons 4, 5 and 6, Quiz Form B
– Continued –

3. The simple interest earned (I) by a savings deposit in a given time, varies jointly as the principal (P) and the interest rate (R). If the interest is \$200 when the principal is \$1,500 and the interest rate is 0.05, find the interest for the same amount of time when the principal is \$900, and the rate is still 0.05.

Quiz Form A Name ____________________

Class __________ Date __________ Score ________

Unit VII – Relations of Rational Number Degree
Part A – Rational Numbers as Exponents
Lesson 1 – Fractions as Exponents
Lesson 2 – Odd and Even "kth" Roots

Use exponents to rewrite each expression in problems 1-8.

1. $5 \times 5 \times 5$ ____________________

2. $3 \cdot 3 \cdot 2 \cdot 2 \cdot 2$ ____________________

3. $x \cdot x \cdot x \cdot y$ ____________________

4. $x \cdot x + 1$ ____________________

5. $a \cdot a - b \cdot b + c \cdot c$ ____________________

6. $\sqrt[5]{\left(^{-}3\right)^5}$ ____________________

7. x^7 __________ · __________
(identical factors)

8. $\sqrt[5]{x^7}$ ____________________

For each of the radicals in problems 9-20, find the indicated root or roots, if they exist.

9. $\sqrt[4]{256}$

10. $\sqrt[3]{27}$

11. $\sqrt[2]{169}$

12. $\sqrt[5]{^{-}32}$

13. $-\sqrt[6]{64}$

14. $\sqrt[11]{^{-}1}$

Name ___________________________________

Unit VII, Part A, Lessons 1 and 2, Quiz Form A
– Continued –

15. $-\sqrt[4]{81}$

16. $\sqrt[2]{^{-}4}$

17. $\sqrt[2]{400}$

18. $-\sqrt[3]{^{-}125}$

19. $\sqrt[3]{\frac{125}{27}}$

20. $-\sqrt[2]{\frac{64}{81}}$

Quiz Form B

Name ______________________

Class __________ Date ____________ Score ________

Unit VII – Relations of Rational Number Degree
Part A – Rational Numbers as Exponents
Lesson 1 – Fractions as Exponents
Lesson 2 – Odd and Even "kth" Roots

Use exponents to rewrite each expression in problems 1-8.

1. $7 \times 7 \times 7 \times 7$ ____________

2. $4 \cdot 4 \cdot 4 \cdot 3 \cdot 3$ ____________

3. $x \cdot x \cdot y \cdot y \cdot y$ ____________

4. $x \cdot x \cdot x + 3$ ____________

5. $\sqrt[7]{(^{-}2)^7}$ ____________

6. a^5 ________ · ________ (identical factors)

7. $x \cdot x \cdot x + y \cdot y - z \cdot z \cdot z \cdot z$ ____________

8. $\sqrt[7]{x^9}$ ____________

For each of the radicals in problems 9-20, find the indicated root or roots, if they exist.

9. $\sqrt[4]{625}$

10. $\sqrt[3]{512}$

11. $\sqrt[2]{169}$

12. $-\sqrt[3]{^{-}8}$

13. $-\sqrt[2]{256}$

14. $\sqrt[5]{^{-}243}$

Name ________________________________

Unit VII, Part A, Lessons 1 and 2, Quiz Form B
– Continued –

15. $\sqrt[2]{^-25}$

16. $\sqrt[15]{^-1}$

17. $\sqrt[2]{441}$

18. $-\sqrt[3]{343}$

19. $\sqrt[4]{\frac{16}{81}}$

20. $-\sqrt[2]{\frac{36}{49}}$

Quiz Form A

Name ____________________

Class ________ Date ________ Score ______

Unit VII – Relations of Rational Number Degree
Part B – Operations with Radical Expressions

Lesson 1 – Multiplication
Lesson 2 – Simplifying with Perfect Powers

For problems 1-6, find the indicated product. Then simplify your answer completely.

1. $\sqrt[2]{8} \cdot \sqrt[2]{20}$

2. $\sqrt[3]{9} \cdot \sqrt[3]{6}$

3. $\sqrt[3]{a^3} \cdot \sqrt[4]{a^3} \cdot \sqrt[2]{a}$

4. $\sqrt[4]{a^5} \cdot \sqrt[4]{a^7}$

5. $\sqrt[2]{5x^5} \cdot \sqrt[2]{12y^3}$

6. $\sqrt[4]{9} \cdot \sqrt[2]{5}$

Simplify each of the radicals in problems 7-14, finding all perfect powers as indicated by the index, and making sure the order of the radical is reduced.

7. $\sqrt{45}$

8. $\sqrt{16a^{10}}$

9. $\sqrt{27e^4 f^3}$

10. $\sqrt[4]{x^9}$

Name ____________________

Unit VII, Part B, Lessons 1 and 2, Quiz Form A
– Continued –

11. $\sqrt[5]{32c^{11}}$

12. $\sqrt[3]{^{-}27x^{6}y^{9}}$

13. $\sqrt[6]{512x^{9}y^{3}}$

14. $\sqrt[10]{32a^{5}}$

Quiz Form B

Name ______________________________

Class ________ **Date** ________ **Score** ______

Unit VII – Relations of Rational Number Degree
Part B – Operations with Radical Expressions

Lesson 1 – Multiplication
Lesson 2 – Simplifying with Perfect Powers

For problems 1-6, find the indicated product. Then simplify your answer completely.

1. $\sqrt[2]{6} \cdot \sqrt[2]{8}$

2. $\sqrt[3]{x^4} \cdot \sqrt[3]{x^2} \cdot \sqrt[2]{x}$

3. $\sqrt[3]{27} \cdot \sqrt[3]{4}$

4. $\sqrt[5]{y^8} \cdot \sqrt[5]{y^7}$

5. $\sqrt[2]{9x^7} \cdot \sqrt[2]{20y^3}$

6. $\sqrt[4]{11} \cdot \sqrt[2]{7}$

Simplify each of the radicals in problems 7-14, finding all perfect powers as indicated by the index, and making sure the order of the radical is reduced.

7. $\sqrt{72}$

8. $\sqrt[2]{36y^8}$

9. $\sqrt[5]{32x^{13}}$

10. $\sqrt[3]{^{-}125m^9n^6}$

Name ________________________________

Unit VII, Part B, Lessons 1 and 2, Quiz Form B
– Continued –

11. $\sqrt{243c^4d^5}$

12. $\sqrt[5]{a^{11}}$

13. $\sqrt[8]{16a^{16}b^{12}}$

14. $\sqrt[6]{8y^9}$

Quiz Form A

Name ______________________________

Class ________ Date __________ Score ______

Unit VII – Relations of Rational Number Degree
Part B – Operations with Radical Expressions
Lesson 3 – Division and Simplifying

Simplify each of the radical expressions in problems 1-10, being sure there are no fractions left under a radical, and no radicals left in the denominator of any fraction.

1. $\frac{\sqrt{3}}{\sqrt{5}}$

2. $\frac{14}{\sqrt{3}}$

3. $\frac{7\sqrt{2}}{\sqrt{6}}$

4. $\frac{\sqrt{64}}{\sqrt{x^2}}$

5. $\frac{\sqrt{14}}{\sqrt{21a}}$

6. $\sqrt{\frac{35}{40}}$

Name ______________________________

Unit VII, Part B, Lesson 3, Quiz Form A
– Continued –

7. $\dfrac{\sqrt{75}}{\sqrt{50a^7}}$

8. $\dfrac{16}{\sqrt{20a^7}}$

9. $\dfrac{7}{\sqrt[3]{2}}$

10. $\dfrac{24xy}{\sqrt{x^6y^3}}$

Quiz Form B

Name ____________________

Class ________ Date __________ Score ______

Unit VII – Relations of Rational Number Degree
Part B – Operations with Radical Expressions

Lesson 3 – Division and Simplifying

Simplify each of the radical expressions in problems 1-10, being sure there are no fractions left under a radical, and no radicals left in the denominator of any fraction.

1. $\dfrac{\sqrt{2}}{\sqrt{7}}$

2. $\dfrac{17}{\sqrt{3}}$

3. $\dfrac{8\sqrt{3}}{\sqrt{6}}$

4. $\dfrac{\sqrt{81}}{\sqrt{4y^2}}$

5. $\dfrac{\sqrt{22}}{\sqrt{33x}}$

6. $\sqrt{\dfrac{75}{80}}$

Name ________________________________

Unit VII, Part B, Lesson 3, Quiz Form B
– Continued –

7. $\dfrac{\sqrt{48}}{\sqrt{16x^3}}$

8. $\dfrac{^{-}81}{\sqrt{3a^3}}$

9. $\dfrac{11}{\sqrt[4]{3}}$

10. $\dfrac{24ab}{\sqrt{a^4b^3}}$

Quiz Form A

Name ____________________

Class ________ Date ________ Score ______

Unit VII – Relations of Rational Number Degree
Part B – Operations with Radical Expressions
Lesson 4 – Addition and Subtraction

For each of the radical expressions in problems 1-10, find the indicated sums and differences, being sure to simplify your results. For convenience, assume that all place holders represent non-negative real numbers.

1. $8\sqrt{5}+9\sqrt{5}$

2. $18\sqrt{2}-7\sqrt{2}+12\sqrt{2}$

3. $10\sqrt{5}-2\sqrt{5}-9\sqrt{5}$

4. $7\sqrt{8}-\sqrt{18}$

5. $^{-}3\sqrt{20}+2\sqrt{45}-\sqrt{7}$

6. $6\sqrt{11}+\sqrt{99}+2\sqrt{44}$

Name ______________________________

Unit VII, Part B, Lesson 4, Quiz Form A
– Continued –

7. $8\sqrt{a} - 5\sqrt{a}$

8. $^{-}16\sqrt{ab} + 5\sqrt{ab} - \sqrt{ab}$

9. $^{-}2\sqrt{12xy} + 8\sqrt{27xy} - \sqrt{3xy}$

10. $\sqrt{18x^4y} + 3x^2\sqrt{2y}$

Quiz Form B

Name ______________________

Class ________ Date __________ Score ______

Unit VII – Relations of Rational Number Degree
Part B – Operations with Radical Expressions
Lesson 4 – Addition and Subtraction

For each of the radical expressions in problems 1-10, find the indicated sums and differences, being sure to simplify your results. For convenience, assume that all place holders represent non-negative real numbers.

1. $4\sqrt{3}+6\sqrt{3}$

2. $12\sqrt{3}-7\sqrt{3}+2\sqrt{3}$

3. $^{-}4\sqrt{7}+5\sqrt{7}+3\sqrt{7}$

4. $4\sqrt{75}+6\sqrt{27}$

5. $7\sqrt{18}-2\sqrt{50}-\sqrt{12}$

6. $^{-}7\sqrt{98}+6\sqrt{18}-\sqrt{32}$

Name ______________________________

Unit VII, Part B, Lesson 4, Quiz Form B
– Continued –

7. $7\sqrt{x} - 3\sqrt{x} + \sqrt{x}$

8. $4\sqrt{xy} - 2\sqrt{xy} - \sqrt{xy} + 10\sqrt{xy}$

9. $3\sqrt{25cd} - 2\sqrt{81xy} + 4\sqrt{36cd}$

10. $5y\sqrt{125x^2} + 8\sqrt{80x^3y^3}$

Quiz Form A

Name ______________________

Class ________ Date __________ Score ______

Unit VII – Relations of Rational Number Degree
Part B – Operations with Radical Expressions

Lesson 5 – Radical Expressions in Polynomials
Lesson 6 – Rationalizing Denominators

In problems 1-6, find each indicated product, being sure to simplify your results. For convenience, assume that all placeholders represent non-negative real numbers.

1. $4\sqrt{7}\cdot 5\sqrt{3}$

2. $10\sqrt{2x}\cdot 3\sqrt{2x}$

3. $7\sqrt{8y^9}\cdot\sqrt{2y^5}$

4. $^{-}7\sqrt{2}\left(5\sqrt{3}-2\sqrt{6}\right)$

5. $\left(3\sqrt{5}+\sqrt{2}\right)\left(\sqrt{5}-2\sqrt{2}\right)$

6. $\sqrt[3]{5}\left(4\sqrt[3]{2}-\sqrt[3]{11}\right)$

In problems 7-12, rationalize each denominator, being sure that your result is simplified.

7. $\sqrt{\frac{3}{5}}$

8. $\frac{3-\sqrt{7}}{\sqrt{6}}$

Name ______________________________

Unit VII, Part B, Lessons 5 and 6, Quiz Form A
– Continued –

9. $\dfrac{{}^{-}10}{\sqrt{y^5}}$

10. $\dfrac{{}^{-}12}{\sqrt{7}-4}$

11. $\dfrac{30x}{\sqrt[3]{4x^2}}$

12. $\dfrac{5}{3\sqrt{5}-1}$

Quiz Form B

Name ______________________

Class __________ Date ____________ Score ________

Unit VII – Relations of Rational Number Degree
Part B – Operations with Radical Expressions
Lesson 5 – Radical Expressions in Polynomials
Lesson 6 – Rationalizing Denominators

In problems 1-6, find each indicated product, being sure to simplify your results. For convenience, assume that all policyholders represent non-negative real numbers.

1. $10\sqrt{5} \cdot 3\sqrt{2}$

2. $5\sqrt{3y} \cdot 7\sqrt{6y}$

3. $^{-}2\sqrt{2a^{10}} \cdot 4\sqrt{12a^{6}}$

4. $4\sqrt{2}\left(5\sqrt{6} - 3\sqrt{2}\right)$

5. $\left(3\sqrt{6} + \sqrt{5}\right)\left(2\sqrt{6} - \sqrt{5}\right)$

6. $\sqrt[5]{7}\left(3\sqrt[5]{3} - \sqrt[5]{9}\right)$

In problems 7-12, rationalize each denominator, being sure that your result is simplified.

7. $\sqrt{\frac{5}{7}}$

8. $\frac{5+\sqrt{6}}{\sqrt{7}}$

Name ______________________________

Unit VII, Part B, Lessons 5 and 6, Quiz Form B
– Continued –

9. $\frac{^-11}{\sqrt{x^3}}$

10. $\frac{^-14}{\sqrt{11}-5}$

11. $\frac{20y}{\sqrt[5]{3^2y^3}}$

12. $\frac{^-11}{4\sqrt{7}+2}$

Quiz Form A

Name ______________________

Class ________ Date ________ Score ______

Unit VII – Relations of Rational Number Degree
Part C – Solving Radical Equations
Lesson 1 – Equations with One Radical Expression
Lesson 2 – Equations with Two Radicals or More

Solve each of the equations in problems 1-4 by isolating the radical (one at a time, if more than one) and applying the principle of powers. Be sure to check your apparent solutions for extraneous roots, and then give the solution set.

1. $9+\sqrt{x-1}=1$

2. $\sqrt{y+5}=\sqrt{2y-3}$

Name ______________________________

Unit VII, Part C, Lessons 1 and 2, Quiz Form A
– Continued –

3. $\sqrt{n+12}-\sqrt{n}=2$

4. $\sqrt{y-5}-\sqrt{y}=1$

Quiz Form B Name ______________________

Class __________ Date __________ Score ______

Unit VII – Relations of Rational Number Degree
Part C – Solving Radical Equations
Lesson 1 – Equations with One Radical Expression
Lesson 2 – Equations with Two Radicals or More

Solve each of the equations in problems 1-4 by isolating the radical (one at a time, if more than one) and applying the principle of powers. Be sure to check your apparent solutions for extraneous roots, and then give the solution set.

1. $5-\sqrt{b+2}=0$

2. $\sqrt{4x-2}=\sqrt{3x+7}$

Name ______________________________

Unit VII, Part C, Lessons 1 and 2, Quiz Form B
– Continued –

3. $\sqrt{c+4}=\sqrt{c+20}-2$

4. $\sqrt{3x+1}+2\sqrt{x}=0$

Quiz Form A

Name ____________________

Class ________ Date ________ Score ______

Unit VII – Relations of Rational Number Degree
Part D – Problem-Solving with Relations Containing Radicals
Lesson 1 – The "Distance" Relation
Lesson 2 – "Formula" Problems

For problems 1 and 2, find the distance between the given pairs of points.

1. (2, 6) and (⁻4, ⁻2)

2. (⁻2, 1) and (4, 2)

Solve problems 3 and 4 using the Pythagorean Theorem. You may leave solutions in radical form, but simplify them completely.

3. A carpenter wants to build a diagonal brace for a gate that is 4 feet wide and 5 feet high. Find the length of the brace.

Name ________________________________

Unit VII, Part D, Lessons 1 and 2, Quiz Form A
– Continued –

4. A rectangular swimming pool is 25 meters long and 10 meters wide. What is the diagonal length of the pool?

Quiz Form B

Name ______________________________

Class __________ Date ___________ Score _______

Unit VII – Relations of Rational Number Degree
Part D – Problem-Solving with Relations Containing Radicals

Lesson 1 – The "Distance" Relation
Lesson 2 – "Formula" Problems

For problems 1 and 2, find the distance between the given pairs of points.

1. (6, 2) and ($^-$5, 2)

2. (2, 22) and ($^-$8, 1)

Solve problems 3 and 4 using the Pythagorean Theorem. You may leave solutions in radical form, but simplify them completely.

3. A 6-meter ladder is leaning against a building. Its base is 1 meter from the base of the building. How high up the building will the ladder reach?

Name ______________________________

Unit VII, Part D, Lessons 1 and 2, Quiz Form B
– Continued –

4. Find the length of the diagonal across a television screen that is 16 cm wide and 12 cm long.

Quiz Form A

Name ____________________

Class ________ Date __________ Score ______

Unit VII – Relations of Rational Number Degree
Part E – The Complex Numbers as a Mathematical System

Lesson 1 – Imaginary and Complex Numbers
Lesson 2 – Addition and Subtraction
Lesson 3 – Multiplication
Lesson 4 – Division

Simplify each of the expressions in problems 1-8. Be sure to write any complex numbers in the standard form $a + bi$.

1. $\sqrt{^{-}80}$

2. i^{39}

3. $\sqrt{^{-}6} \cdot \sqrt{^{-}12}$

4. $(9+7i)+(15-8i)$

Name ____________________________________

Unit VII, Part E, Lessons 1, 2, 3 and 4, Quiz Form A
– Continued –

5. $(6-9i)-(17-13i)$

6. $(7-4i)(8+3i)$

7. $\dfrac{3+i}{2-i}$

8. $\dfrac{4-i\sqrt{3}}{4+i\sqrt{3}}$

Quiz Form B

Name ____________________

Class ________ Date ________ Score ______

Unit VII – Relations of Rational Number Degree
Part E – The Complex Numbers as a Mathematical System
Lesson 1 – Imaginary and Complex Numbers
Lesson 2 – Addition and Subtraction
Lesson 3 – Multiplication
Lesson 4 – Division

Simplify each of the expressions in problems 1-8. Be sure to write any complex numbers in the standard form $a + bi$.

1. $\sqrt{^{-}75}$

2. i^{29}

3. $\sqrt{^{-}10} \cdot \sqrt{^{-}18}$

4. $(8 + 3i) + (17 - 9i)$

Name ______________________________

Unit VII, Part E, Lessons 1, 2, 3 and 4, Quiz Form B
– Continued –

5. $(4-11i)-(13+3i)$

6. $(6-5i)(3+4i)$

7. $\frac{5+i}{3-i}$

8. $\frac{5-i\sqrt{3}}{5+i\sqrt{3}}$

Quiz Form A Name ______________________

Class __________ Date __________ Score ______

Unit VIII – Quadratic Equations
Part A – Solving Quadratic Equations of the Form $ax^2 + bx + c = 0$
Lesson 1 – Suppose a = 0, b = 0, or c = 0
Lesson 2 – Suppose a, b, c $\neq$ 0

For each of the relations in problems 1-4, identify a, b, and c, relative to the standard form. Then solve using appropriate strategies, and give the solution set.

1. $x^2 - 9 = 0$

a = __________

b = __________

c = __________

2. $x^2 + 5x = 0$

a = __________

b = __________

c = __________

3. $2x = 6$

a = __________

b = __________

c = __________

4. $x^2 - 12x + 27 = 0$

a = __________

b = __________

c = __________

In problems 5 and 6, find the value of "c" that makes each trinomial a perfect square.

5. $x^2 + 14x + c$

6. $y^2 - 5y + c$

Name ___________________________________

Unit VIII, Part A, Lessons 1 and 2, Quiz Form A
– Continued –

Solve each of the relations in problems 7 and 8 by completing the square. Then give the solution set.

7. $x^2 + 8x + 3 = 0$

8. $2x^2 - 4x - 3 = 0$

Quiz Form B

Name ______________________

Class ________ **Date** __________ **Score** ______

Unit VIII – Quadratic Equations
Part A – Solving Quadratic Equations of the Form $ax^2 + bx + c = 0$
Lesson 1 – Suppose a = 0, b = 0, or c = 0
Lesson 2 – Suppose a, b, c ≠ 0

For each of the relations in problems 1-4, identify a, b, and c, relative to the standard form. Then solve using appropriate strategies, and give the solution set.

1. $n^2 - 6n = 0$

a = __________

b = __________

c = __________

2. $m^2 - 8m + 12 = 0$

a = __________

b = __________

c = __________

3. $u^2 - 121 = 0$

a = __________

b = __________

c = __________

4. $4y = 12$

a = __________

b = __________

c = __________

In problems 5 and 6, find the value of "c" that makes each trinomial a perfect square.

5. $m^2 + 10m + c$

6. $x^2 - 7x + c$

Name ___________________________________

Unit VIII, Part A, Lessons 1 and 2, Quiz Form B
– Continued –

Solve each of the relations in problems 7 and 8 by completing the square. Then give the solution set.

7. $x^2 + 4x - 9 = 0$

8. $3x^2 + 4x - 7 = 0$

Quiz Form A

Name ______________________

Class __________ Date __________ Score ________

Unit VIII – Quadratic Equations
Part A – Solving Quadratic Equations of the Form $ax^2 + bx + c = 0$
Lesson 3 – The Quadratic Formula
Lesson 4 – Checking Solutions

For each of the relations in problems 1 and 2, identify a, b, and c, relative to the standard form. Then solve by substituting in the quadratic formula, and give the solution set.

1. $x^2 - 3x - 28 = 0$

a = __________

b = __________

c = __________

2. $2m^2 + 4m - 30 = 0$

a = __________

b = __________

c = __________

Name ______________________________

Unit VIII, Part A, Lessons 3 and 4, Quiz Form A
– Continued –

In each of the relations in problems 3 and 4, find the value of the discriminant. Then describe the roots completely.

3. $7x^2 - 3x + 5 = 0$

4. $3x^2 - 5x - 4 = 0$

Quiz Form B

Name ______________________

Class __________ Date __________ Score ________

Unit VIII – Quadratic Equations
Part A – Solving Quadratic Equations of the Form $ax^2 + bx + c = 0$

Lesson 3 – The Quadratic Formula
Lesson 4 – Checking Solutions

For each of the relations in problems 1 and 2, identify a, b, and c, relative to the standard form. Then solve by substituting in the quadratic formula, and give the solution set.

1. $7x^2 + 3x - 2 = 0$

a = __________

b = __________

c = __________

2. $2x^2 - 6x - 3 = 0$

a = __________

b = __________

c = __________

Name ___________________________________

Unit VIII, Part A, Lessons 3 and 4, Quiz Form B
– Continued –

In each of the relations in problems 3 and 4, find the value of the discriminant. Then describe the roots completely.

3. $9t^2 - 12t + 4 = 0$

4. $20u^2 + 13u + 2 = 0$

Quiz Form A

Name ______________________

Class __________ Date __________ Score ______

Unit VIII – Quadratic Equations
Part A – Solving Quadratic Equations of the Form $ax^2 + bx + c = 0$
Lesson 5 – Quadratic Inequalities

Solve the following quadratic inequality by finding the critical points and determining which interval(s) satisfy the relation. Show your solution, first by a graph on a number line, then by using set notation.

$$2x^2 - x - 6 < 0$$

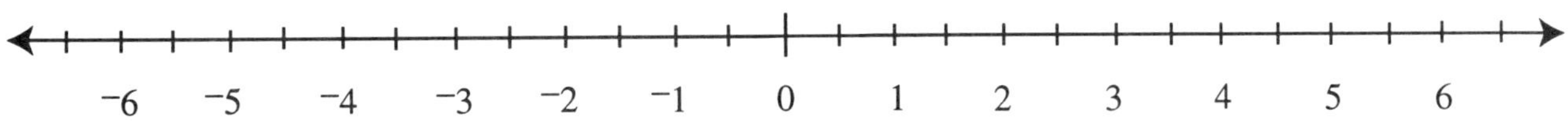

Quiz Form B

Name ____________________

Class ________ Date ________ Score ______

Unit VIII – Quadratic Equations
Part A – Solving Quadratic Equations of the Form $ax^2 + bx + c = 0$
Lesson 5 – Quadratic Inequalities

Solve the following quadratic inequality by finding the critical points and determining which interval(s) satisfy the relation. Show your solution, first by a graph on a number line, then by using set notation.

$$p^2 - 4p - 5 \geq 0$$

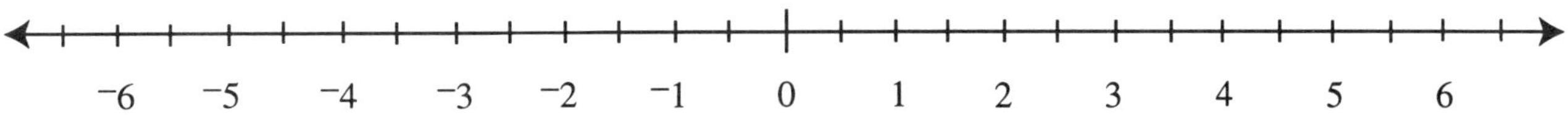

Quiz Form A

Name ______________________

Class ________ **Date** ________ **Score** ______

Unit VIII – Quadratic Equations
Part B – Equations That Are Quadratic in Form
Lesson 1 – Higher Integer Order
Lesson 2 – Lower Rational Order, Greater Then Zero

Solve each of the equations in problems 1 and 2 by using a temporary substitution to "reduce" it to a quadratic equation in standard form. Then give the solution set for the original relation, being sure to check for extraneous roots.

1. $x^{\frac{2}{3}} - 12x^{\frac{1}{3}} + 20 = 0$

Name ________________________________

Unit VIII, Part B, Lessons 1 and 2, Quiz Form A
– Continued –

2. $x^4 - 13x^2 + 36 = 0$

Quiz Form B

Name ______________________

Class ________ Date __________ Score ______

Unit VIII – Quadratic Equations
Part B – Equations That Are Quadratic in Form
Lesson 1 – Higher Integer Order
Lesson 2 – Lower Rational Order, Greater Then Zero

Solve each of the equations in problems 1 and 2 by using a temporary substitution to "reduce" it to a quadratic equation in standard form. Then give the solution set for the original relation, being sure to check for extraneous roots.

1. $x^4 - 7x^2 + 12 = 0$

Name ________________________________

Unit VIII, Part B, Lessons 1 and 2, Quiz Form B
– Continued –

2. $x^{\frac{1}{2}} - 6x^{\frac{1}{4}} + 8 = 0$

Quiz Form A

Name ______________________

Class __________ Date __________ Score ________

Unit VIII – Quadratic Equations
Part B – Equations That Are Quadratic in Form
Lesson 3 – Integer Order, Less Than Zero

Solve each of the equations in problems 1 and 2 by using a temporary substitution to "reduce" it to a quadratic equation in standard form. Then give the solution set for the original relation, being sure to check for extraneous roots.

1. $y^{-1} - 8y^{-\frac{1}{2}} + 12 = 0$

Name ____________________________________

Unit VIII, Part B, Lesson 3, Quiz Form A
– Continued –

2. $2x^{-2} + x^{-1} - 1 = 0$

Quiz Form B

Name ______________________

Class ________ Date ________ Score ______

Unit VIII – Quadratic Equations
Part B – Equations That Are Quadratic in Form
Lesson 3 – Integer Order, Less Than Zero

Solve each of the equations in problems 1 and 2 by using a temporary substitution to "reduce" it to a quadratic equation in standard form. Then give the solution set for the original relation, being sure to check for extraneous roots.

1. $m^{-2} + 9m^{-1} - 10 = 0$

Name ___________________________________

Unit VIII, Part B, Lesson 3, Quiz Form B
– Continued –

2. $3x^{-2} - x^{-1} - 14 = 0$

Quiz Form A

Name ______________________

Class ________ Date ________ Score ______

Unit VIII – Quadratic Equations
Part C – Problem Solving With Quadratic Relations
Lesson 1 – "Geometric Figure" Problems
Lesson 2 – "Pythagorean Theorem" Problems

For both of the following story problems, answer the four analysis questions to find the system of equations needed to solve. Then solve, and use common sense to check your answer.

1. A 20-foot ladder leans against a building with the base of the ladder a specified distance "d" from the building. The ladder will reach 4 feet higher on the side of the building than the specified distance "d." Find "d" and the distance to the top of the ladder.

Name ______________________________

Unit VIII, Part C, Lessons 1 and 2, Quiz Form A
– Continued –

2. The width of a rectangular lot is 8 meters less than the length. The area of the lot is 20 square meters. Find the length and the width.

Quiz Form B

Name ____________________

Class ________ Date ________ Score ______

Unit VIII – Quadratic Equations
Part C – Problem Solving With Quadratic Relations
Lesson 1 – "Geometric Figure" Problems
Lesson 2 – "Pythagorean Theorem" Problems

For both of the following story problems, answer the four analysis questions to find the system of equations needed to solve. Then solve, and use common sense to check your answer.

1. The base of a triangle is 2 more than twice the height of the triangle. The area of the triangle is 12 square centimeters. Find the height of the triangle and the length of the base of the triangle.

Name ______________________________

Unit VIII, Part C, Lessons 1 and 2, Quiz Form B
– Continued –

2. The hypotenuse of a right triangle is 26 feet long. The length of one leg is 14 feet more than the other. Find the lengths of the legs.

Quiz Form A

Name ______________________

Class ________ Date __________ Score ______

Unit VIII – Quadratic Equations
Part C – Problem Solving With Quadratic Relations

Lesson 3 – "Work" Problems
Lesson 4 – "Motion" Problems

For both of the following story problems, answer the four analysis questions to find the equation needed to solve, and use common sense to check your answer.

1. A car travels 200 miles at a certain speed. If the speed had been 10 mph faster, the trip would have taken one hour less. Find the speed.

Name ________________________________

Unit VIII, Part C, Lessons 3 and 4, Quiz Form A
– Continued –

2. Two people can do a job in 4 hours. Person "A" requires 6 hours longer, working alone, than person "B." How long would it take person "B" to do the same job alone?

Quiz Form B

Name ______________________

Class ________ Date __________ Score ______

Unit VIII – Quadratic Equations
Part C – Problem Solving With Quadratic Relations
Lesson 3 – "Work" Problems
Lesson 4 – "Motion" Problems

For both of the following story problems, answer the four analysis questions to find the equation needed to solve, and use common sense to check your answer.

1. Airplane "A" travels 600 miles at a certain speed. Plane "B" travels 1,000 miles at a speed that is 50 mph faster, but it takes only one hour longer. Find the speed of each plane.

Name ___________________________

Unit VIII, Part C, Lessons 3 and 4, Quiz Form B
– Continued –

2. Two pumps can fill a tank in $1\frac{1}{2}$ hours. One pump requires 4 hours longer, running alone, to fill the tank than the other. How long would it take the faster pump, working alone, to fill the tank?

Unit Test Form A

Name ______________________

Class __________ Date __________ Score ________

Unit VI – Second Degree Relations and Higher – Algebraic Fractions

For each of the rational expressions in problems 1 and 2, find all numbers for which the expression is undefined. In other words, find the restricted values.

1. $\dfrac{x^2 + x + 105}{5x - 45}$

2. $\dfrac{x^2 - 3x - 4}{x^2 - 18x + 77}$

Simplify each of the rational expressions in problems 3-5 by factoring to make as many ones as possible.

3. $\dfrac{7x^3}{28x^2}$

4. $\dfrac{8x + 24}{8x - 24}$

5. $\dfrac{a^2 + 8a - 9}{a^2 - 5a + 4}$

Perform the indicated operations and simplify in problems 6 and 7.

6. $\dfrac{2x^2 + 3x - 5}{x^2 - 9} \cdot \dfrac{x + 3}{2x + 5}$

7. $\dfrac{m^2 - 9}{5y - 5} \div \dfrac{y - 3}{y - 1}$

8. Find the least common multiple for $x^2 + 6x - 7$ and $x^2 + x - 2$.

Name ____________________________

Unit VI – Second Degree Relations and Higher – Algebraic Fractions
Unit Test Form A
– Continued, Page 2 –

Perform the indicated operations and simplify in problems 9-13.

9. $\dfrac{x-2y}{x+y}+\dfrac{x+9y}{x+y}$

10. $\dfrac{4}{x}-\dfrac{9}{^{-}x}$

11. $\dfrac{2}{x+4}-\dfrac{2x-5}{x^2-16}+\dfrac{3}{x-4}$

12. $\dfrac{2a}{a^2-c^2}-\dfrac{1}{a+c}+\dfrac{1}{a-c}$

13. $\dfrac{9}{3-y}+\dfrac{y^2}{y-3}$

Simplify each of the complex fractions in problems 14 and 15.

14. $\dfrac{\dfrac{x+y}{xy}}{\dfrac{x^3+y^3}{x^3y^3}}$

15. $\dfrac{\dfrac{1}{x}+\dfrac{1}{y}}{\dfrac{1}{x^2}-\dfrac{1}{y^2}}$

Name ______________________________

Unit VI – Second Degree Relations and Higher – Algebraic Fractions
Unit Test Form A
– Continued, Page 3 –

Solve each of the relations in problems 16-21, and show the solution sets of the inequalities on the accompanying number lines.

16. $\frac{x}{3} - \frac{x}{4} = 12$

17. $\frac{3}{8} + \frac{1}{3} = \frac{m}{12}$

18. $\frac{4}{x-7} = \frac{2}{x+5}$

19. $\frac{1}{3x-6} - \frac{1}{x^2-4} = \frac{3}{x+2}$

20. $\frac{x}{x-5} < 2$

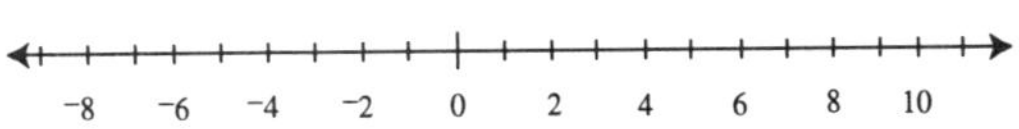

21. $\frac{x+1}{x-2} \geq 0$

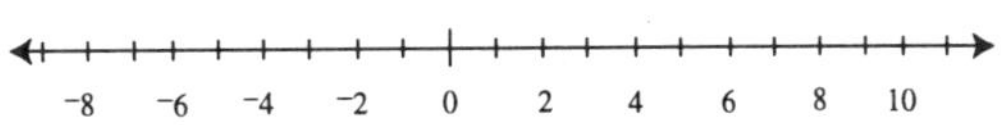

Name ______________________________

Unit VI – Second Degree Relations and Higher – Algebraic Fractions
Unit Test Form A
– Continued, Page 4 –

Solve each of the relations in problems 22 and 23 for the indicated placeholder.

22. $\frac{W_1}{W_2} = \frac{d_1}{d_2}$ for d_1

23. $\frac{1}{x} + \frac{1}{y} = \frac{1}{w}$ for y

24. An airplane flies 1,200 miles with the wind. In the same amount of time, it can fly 800 miles against the wind. The speed of the plane in still air is 250 miles per hour. Find the speed of the wind.

25. A tank can be filled in 10 hours by pump "A" working alone. Working together, it takes pump "A" and pump "B" 6 hours to do the same job. How long would it take pump "B" to fill the tank by itself?

26. Explain the difference between a rational expression and a rational equation. The phrase "clearing fractions" would apply to which one? Explain.

Unit Test Form B

Name ______________________

Class ________ Date ________ Score ______

Unit VI – Second Degree Relations and Higher – Algebraic Fractions

For each of the rational expressions in problems 1 and 2, find all numbers for which the expression is undefined. In other words, find the restricted values.

1. $\dfrac{a^4 - 5a}{a^2 - 3a - 28}$

2. $\dfrac{x-6}{x+8}$

Simplify each of the rational expressions in problems 3-5 by factoring to make as many ones as possible.

3. $\dfrac{48m^5}{56m^{11}}$

4. $\dfrac{12x-15}{21}$

5. $\dfrac{x^2 - 9x + 8}{x^2 + 3x - 4}$

Perform the indicated operations and simplify in problems 6 and 7.

6. $\dfrac{y^2 + 10y + 25}{y^2 - 9} \cdot \dfrac{y^2 - 3y}{y+5}$

7. $\dfrac{a^2 - 36}{a^2 - 8a + 16} \div \dfrac{3a - 18}{a^2 - a - 12}$

8. Find the least common multiple for $x^2 - y^2$ and $3y - 3x$.

Name ______________________________

Unit VI – Second Degree Relations and Higher – Algebraic Fractions
Unit Test Form B
– Continued, Page 2 –

Perform the indicated operations and simplify in problems 9-13.

9. $\dfrac{y-2}{4y+8}-\dfrac{y+6}{5y+10}$

10. $\dfrac{a^2}{a-b}+\dfrac{b^2}{b-a}$

11. $\dfrac{2}{y+3}-\dfrac{y}{y-1}+\dfrac{y^2+2}{y^2+2y-3}$

12. $\dfrac{4x}{x^2-1}+\dfrac{3x}{x+1}-\dfrac{4}{x-1}$

13. $\dfrac{5ab}{a^2-b^2}+\dfrac{a+b}{a-b}$

Simplify each of the complex fractions in problems 14 and 15.

14. $\dfrac{1-\dfrac{1}{x^2}}{1-\dfrac{1}{x}}$

15. $\dfrac{\dfrac{1}{x}+\dfrac{2}{y}}{\dfrac{3}{x}-\dfrac{1}{y}}$

Name ______________________________

Unit VI – Second Degree Relations and Higher – Algebraic Fractions
Unit Test Form B
– Continued, Page 3 –

Solve each of the relations in problems 16-21, and show the solution sets of the inequalities on the accompanying number lines.

16. $\frac{x}{10}=\frac{2}{5}+\frac{3}{8}$

17. $x+\frac{8}{x}={}^{-}9$

18. $\frac{3}{y+1}=\frac{2}{y-3}$

19. $\frac{2}{x+5}+\frac{1}{x-5}=\frac{16}{x^2-25}$

20. $\frac{x+1}{x-2}\geq 3$

21. $\frac{x-2}{x+5}<0$

Name ____________________

Unit VI – Second Degree Relations and Higher – Algebraic Fractions
Unit Test Form B
– Continued, Page 4 –

Solve each of the relations in problems 22 and 23 for the indicated placeholder.

22. $\frac{t}{a} + \frac{t}{b} = 1$ for b

23. $s = \frac{(v_1 + v_2)t}{2}$ for t

24. The electric current I in a circuit varies directly as the voltage V. When 20 volts are applied, the current is 5 amperes. What is the current when 30 volts are applied?

25. The sum of a number and its reciprocal is $\frac{61}{30}$. Find the number.

26. Explain how you might estimate the number of words in a novel by using an idea studied in this unit.

Cumulative Review Form A

Name ____________________

Class ________ Date ________ Score ______

Algebra: A Complete Course
Units I – VI

Simplify. Write final answers with positive exponents.

1. $^{-}5x^{^{-}4}y^{0}$

2. $\left(\dfrac{12y^{^{-}5}}{15y^{^{-}3}}\right)^{^{-}4}$

Convert to scientific notation.

3. 0.00000717

4. 18,600,000

Simplify, and write the final answer in scientific notation.

5. $\left(8.6\times10^{^{-}8}\right)\times\left(7.3\times10^{^{-}6}\right)$

6. $\dfrac{3.2\times10^{9}}{2.5\times10^{6}}$

Simplify.

7. $6-2(3x-5)$

8. $25-5^{2}+20\div2\cdot(3-8)$

Name ______________________

Cumulative Review Form A
Algebra: A Complete Course, Units I – VI

– Continued, Page 2 –

9. $\dfrac{\frac{1}{x} - \frac{1}{y}}{\frac{1}{xy}}$

10. $\dfrac{\frac{2}{x+y}}{\frac{4}{x+y}}$

Divide and simplify.

11. $\dfrac{k^4}{k^2+3k+2} \div \dfrac{16k^3}{k+2}$

Add and simplify.

12. $\dfrac{2}{x-4} + \dfrac{3x}{x^2-16}$

Multiply.

13. $(5x-2)^2$

14. $(4a+1)(3a^2-2a-1)$

Factor completely.

15. $m^3 - 8n^3$

16. $6x^2 + x - 15$

Name ______________________

Cumulative Review Form A
Algebra: A Complete Course, Units I – VI

– Continued, Page 3 –

Solve.

17. $4+\frac{8}{a-5}=\frac{a}{a-5}$

18. $\frac{2}{p}+\frac{3}{p-1}=\frac{5}{p^2-p}$

19. $4x^3-9x=0$

20. $2x+1\le {}^{-}5$

21. $\frac{8y+12}{4y}=\frac{6}{5}$

22. $\frac{P_1V_1}{T_1}=\frac{P_2V_2}{T_2}$ for P_1

Name ______________________________

Cumulative Review Form A
Algebra: A Complete Course, Units I – VI

– Continued, Page 4 –

23. Through years of diligent saving, Juan has \$25,000 to invest. Juan plans on investing part of the money into an account that pays 5% annually and the rest into an account that pays 10% annually. To supplement his income, Juan wants to earn at least \$2,030 each year. How much must he invest into each account to earn exactly \$2,030?

Cumulative Review Form B Name ______________________

Class __________ Date __________ Score ______

Algebra: A Complete Course
Units I – VI

Simplify. Write final answers with positive exponents.

1. $^-7x^0y^{-3}$

2. $\left(\dfrac{20a^{-5}b^2}{15a^{-4}}\right)^{-3}$

Convert to scientific notation.

3. 0.0000012

4. 758,000

Simplify, and write the final answer in scientific notation.

5. $\left(7.2\times10^{-9}\right)\times\left(8.9\times10^{-7}\right)$

6. $\dfrac{2.5\times10^{-1}}{4\times10^3}$

Solve, and graph the result on a number line.

7. $^-15<8x-3\le13$

8. $2x-7<9$ and $x+4\ge3$

Name ______________________________

Cumulative Review Form B
Algebra: A Complete Course, Units I – VI

– Continued, Page 2 –

9. Graph $x \geq 5y - 2$.

10. Find the equation of a line containing the points $(2, {}^{-}3)$ and (5, 7).

Divide and simplify.

11. $\dfrac{6k^8}{k^2 + k - 6} \div \dfrac{8k^2}{k - 2}$

Add and simplify.

12. $\dfrac{3}{x^2 - 9} + \dfrac{2x}{x + 3}$

Multiply.

13. $(7x - 6)^2$

14. $(5a - 3)(3a^2 - 8a - 1)$

Name ____________________________________

Cumulative Review Form B
Algebra: A Complete Course, Units I – VI

– Continued, Page 3 –

Factor completely.

15. $27x^3 - 125y^3$

16. $8x^2 + 8x - 70$

Solve.

17. $^{-}1 + \dfrac{4}{a-5} = \dfrac{a}{a-5}$

18. $x^3 - 36x = 0$

19. $\dfrac{5}{p-4} - \dfrac{2}{p^2 - 4p} = \dfrac{1}{p}$

20. $\dfrac{6y+1}{8y} = \dfrac{5}{6}$

21. $4x - 3 \le {}^{-}23$

22. $A = \dfrac{KB}{C}$ for C

Name ______________________________

Cumulative Review Form B
Algebra: A Complete Course, Units I – VI

– Continued, Page 4 –

23. The average distance between the earth and the sun is 1.5×10^8 kilometers. If NASA sends a rocket from the earth to the sun that averages 8×10^3 kilometers per hour, how long will it take the rocket to reach the sun.

Unit Test Form A

Name ______________________

Class ________ Date ________ Score ______

Unit VII – Relations of Rational Number Degree

Rewrite each of the expressions in problems 1-4 using rational exponents. Assume that any placeholder may represent any real number.

1. $4 \cdot 4 \cdot 4 \cdot x \cdot x \cdot x \cdot x \cdot y \cdot y$

2. $a \cdot a \cdot a + 2$

3. $\sqrt[3]{\left(^{-}2\right)^3}$

4. $\sqrt[6]{x^8}$

For all remaining problems on this test, assume that all placeholders represent nonnegative numbers.

Simplify each of the radical expressions in problems 5-9.

5. $\sqrt{12x^2yz^3}$

6. $\sqrt{9^3}$

7. $\sqrt[3]{162x^2y^5}$

8. $\sqrt[4]{48a^5b^8}$

Name ______________________________

Unit VII – Relations of Rational Number Degree, Unit Test Form A – Continued, Page 2 –

9. $\sqrt{x^2 + 8x + 16}$

In problems 10-16, perform the indicated operations, and simplify.

10. $\sqrt[3]{x^5} \cdot \sqrt[3]{27xy}$

11. $\sqrt{6x^3y} \cdot \sqrt{3xy^2}$

12. $\dfrac{\sqrt{63a^4b^2}}{\sqrt{7ab}}$

13. $\dfrac{\sqrt{100x^3}}{\sqrt{4x}}$

14. $\sqrt{2(x+1)^3} \cdot \sqrt{4(x+1)}$

15. $\dfrac{\sqrt[3]{8xy^5}}{\sqrt[3]{xy}}$

Name ______________________________

Unit VII – Relations of Rational Number Degree, Unit Test Form A – Continued, Page 3 –

16. $\left(\sqrt[3]{4a^3b}\right)^2$

In problems 17-21, find the indicated sum or difference, and simplify.

17. $5\sqrt{2}+8\sqrt{2}$

18. $7\sqrt{45}-2\sqrt{5}$

19. $\sqrt[3]{27}-5\sqrt[3]{8}$

20. $\sqrt{5a}+2\sqrt{45a^3}$

21. $\sqrt[3]{24x}-\sqrt[3]{3x^4}$

Name ______________________________

Unit VII – Relations of Rational Number Degree, Unit Test Form A – Continued, Page 4 –

For each fraction in problems 22-24, rationalize the denominator. Then simplify your result.

22. $\frac{\sqrt{2}}{\sqrt{5}}$

23. $\frac{3x^2}{\sqrt[3]{2y}}$

24. $\frac{4}{8-\sqrt{5}}$

Solve each of the relations in problems 25-27.

25. $x=\sqrt{x+7}+5$

26. $\sqrt{5x+2}=7$

Name ______________________________

Unit VII – Relations of Rational Number Degree, Unit Test Form A
– Continued, Page 5 –

27. $\sqrt{2x-5} = 1 + \sqrt{x-3}$

28. Find the distance between $(4, 3)$ and $(^-2, 1)$.

Name ___________________________

Unit VII – Relations of Rational Number Degree, Unit Test Form A – Continued, Page 6 –

29. The diagonal of a square has a length of $11\sqrt{2}$. Find the length of a side of the square.

30. A baseball diamond is set up so that the distance from home plate to first base is 45 feet. To the nearest foot, how far is it from home plate to second base?

Name ____________________

Unit VII – Relations of Rational Number Degree, Unit Test Form A – Continued, Page 7 –

31. Determine whether $1+i$ is a solution of $2x^2-4x+4=0$. Explain why or why not.

Simplify each of the expressions in problems 32-38.

32. $\sqrt{^-5}$

33. $\sqrt{^-7}\cdot\sqrt{^-5}$

34. i^{31}

35. $(8+3i)-(^-2+5i)$

Unit VII – Relations of Rational Number Degree, Unit Test Form A – Continued, Page 8 –

36. $(8-3i)^2$

37. $(2+3i)(3-i)$

38. $\dfrac{6+2i}{4-i}$

39. Simplify: $\sqrt{\sqrt{\left(\sqrt{\left(\sqrt{\left(\sqrt{2^2}\right)^2}\right)^4}\right)^2}}$

Unit Test Form B Name ______________________

Class __________ Date __________ Score ________

Unit VII - Relations of Rational Number Degree

Rewrite each of the expressions in problems 1-4 using rational exponents. Assume that any placeholder may represent any real number.

1. $\sqrt[4]{x^6}$

2. $9 \cdot 9 \cdot 9 \cdot 9 \cdot m \cdot m \cdot m \cdot t \cdot t$

3. $\sqrt[5]{\left({}^{-}3\right)^5}$

4. $5 + x \cdot x \cdot x \cdot x$

For all remaining problems on this test, assume that all placeholders represent nonnegative numbers.

Simplify each of the radical expressions in problems 5-9.

5. $\sqrt{18a^3b^2c}$

6. $\sqrt[4]{81^3}$

7. $\sqrt[3]{200x^3y^4}$

8. $\sqrt{x^2 - 10x + 25}$

Name ________________________________

Unit VII – Relations of Rational Number Degree, Unit Test Form B – Continued, Page 2 –

9. $\sqrt{x^2 + 8x + 16}$

In problems 10-16, perform the indicated operations, and simplify.

10. $\sqrt{2x^3y} \cdot \sqrt{12xy}$

11. $\sqrt[3]{x}\left(\sqrt[3]{3x^2} - \sqrt[3]{81x^2}\right)$

12. $\sqrt[5]{\dfrac{32b^{10}}{243a^{20}}}$

13. $\sqrt[3]{2(y+5)}\sqrt[3]{4(y+5)^4}$

14. $\dfrac{\sqrt[4]{48x^9y^{13}}}{\sqrt[4]{3xy^5}}$

15. $\dfrac{\sqrt{x^3 - y^3}}{\sqrt{x - y}}$

Name ________________________________

Unit VII – Relations of Rational Number Degree, Unit Test Form B – Continued, Page 3 –

16. $\left(\sqrt[5]{3x^2y^3}\right)^3$

In problems 17-21, find the indicated sum or difference, and simplify.

17. $4\sqrt{3}+7\sqrt{3}$

18. $8\sqrt{63}-4\sqrt{7}$

19. $3\sqrt[3]{16}+\sqrt[3]{54}$

20. $4\sqrt{3x^3}-\sqrt{12x}$

21. $\sqrt[3]{54x}-\sqrt[3]{2x^4}$

Name ____________________________________

Unit VII – Relations of Rational Number Degree, Unit Test Form B – Continued, Page 4 –

For each fraction in problems 22-24, rationalize the denominator. Then simplify your result.

22. $\dfrac{\sqrt[3]{a}}{\sqrt[3]{9x}}$

23. $\dfrac{5}{3+\sqrt{2}}$

24. $\dfrac{\sqrt{5}}{\sqrt{7}}$

Solve each of the relations in problems 25-27.

25. $\sqrt{x-3}+\sqrt{x+5}=4$

26. $\sqrt{2y+9}=5$

Name ______________________________

Unit VII – Relations of Rational Number Degree, Unit Test Form B – Continued, Page 5 –

27. $\sqrt{y-5}+\sqrt{y}=5$

28. Find the distance between $\left({}^{-}5, 2\right)$ and $\left(3, {}^{-}1\right)$.

Name ___________________________________

Unit VII – Relations of Rational Number Degree, Unit Test Form B – Continued, Page 6 –

29. How long is a guy wire reaching from the top of a 15-foot pole to a point on the ground 10 feet from the base of the pole?

30. A 25-inch (diagonal screen length) TV set has a screen with a height of 15 inches. What is the width?

Name ________________________________

Unit VII – Relations of Rational Number Degree, Unit Test Form B – Continued, Page 7 –

31. Use rational exponents to write a single radical expression: $\sqrt[4]{5x} \cdot \sqrt{y+3}$

Simplify each of the expressions in problems 32-38.

32. $\sqrt{^-2} \cdot \sqrt{^-3}$

33. $\sqrt{^-7}$

34. i^{24}

35. $(7+i)^2$

Name ____________________________________

Unit VII – Relations of Rational Number Degree, Unit Test Form B – Continued, Page 8 –

36. $(7+9i)(^{-}6+4i)$

37. $(7+2i)^2$

38. $\dfrac{3+i}{5-i}$

39. Find: $i \cdot i^2 \cdot i^3 \cdot i^4 \cdot i^5 \cdots\cdots i^{98} \cdot i^{99} \cdot i^{100}$ Then describe the pattern you found.

Unit Test Form A

Name ______________________

Class ________ Date ________ Score ______

Unit VIII – Quadratic Equations

Find the solution set for each of the relations in problems 1-6.

1. $3x^2 + x - 2 = 0$

2. $n^2 - 9 = 0$

3. $x^2 - 5x = 0$

4. $3x - 12 = 0$

5. $5p^2 - 9 = 0$

6. $(w + 2)^2 = 11$

Name ______________________________

Unit VIII – Quadratic Equations, Unit Test Form A
– Continued, Page 2 –

Find the solution set for each of the relations in problems 7-10 by completing the square.

7. $n^2 + 6n + 7 = 0$

8. $x^2 - 6x + 8 = 0$

9. $2v^2 - 3v + 7 = 0$

10. $x^2 + 3x + 1 = 0$

Name ________________________________

Unit VIII – Quadratic Equations, Unit Test Form A
– Continued, Page 3 –

Find the solution set for each of the relations in problems 11 and 12 by using the quadratic formula.

11. $5x^2 - 8x - 3 = 0$

12. $x^2 - x + 2 = 0$

In problems 13 and 14, find the value of the discriminant, and state the nature of the solutions.

13. $4x^2 + 8x - 5 = 0$

14. $x^2 + 3x + 4 = 0$

Name ________________________________

Unit VIII – Quadratic Equations, Unit Test Form A
– Continued, Page 4 –

Show the solution set of each of the relations in problems 15-17 by finding the critical points on a number line and determining which interval(s) make the given relation true. Then write the solution using set notation.

15. $c^2 + 3c > 4$

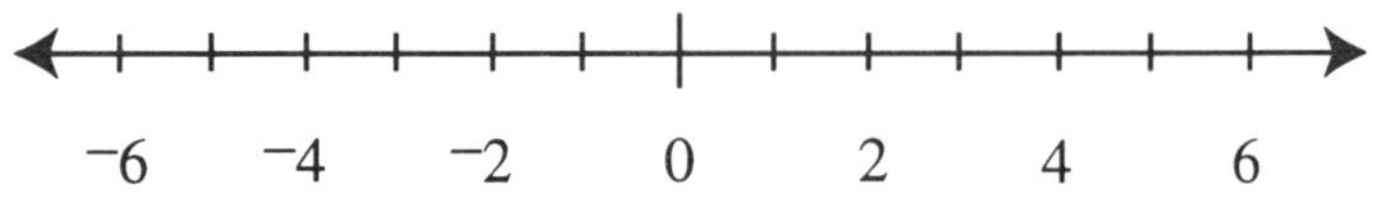

16. $x^2 + 8 \leq 6x$

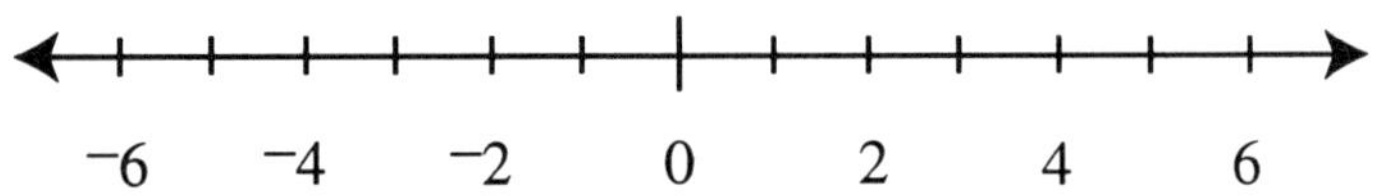

Name ________________________________

Unit VIII – Quadratic Equations, Unit Test Form A
– Continued, Page 5 –

17. $\dfrac{m-1}{m-3} > 0$

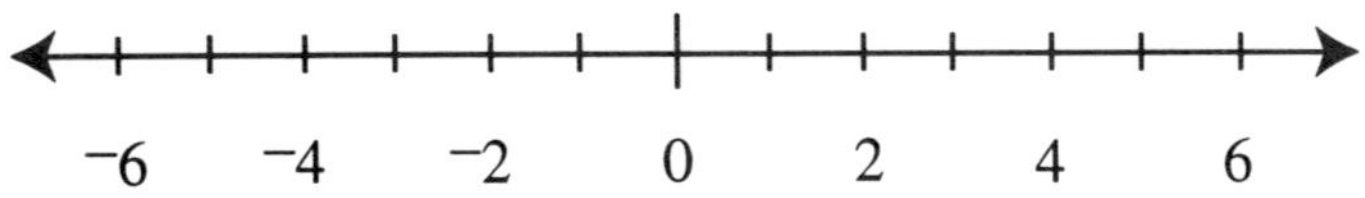

In problems 18 and 19, find the solution set of each relation by using a temporary substitution to "reduce" it to a quadratic equation in standard form.

18. $x^4 - 10x^2 + 9 = 0$

19. $x^{-2} + x^{-1} - 6 = 0$

Name ___________________________________

Unit VIII – Quadratic Equations, Unit Test Form A – Continued, Page 6 –

20. Sue and Karen work as student assistants. Together they can correct a set of exams in $3\frac{3}{5}$ hours. It would take Sue 3 hours more than Karen to do the same job alone. How long would each person need to do the job alone?

Name ________________________________

Unit VIII – Quadratic Equations, Unit Test Form A
– Continued, Page 7 –

21. A car travels 240 miles at a certain speed. If the speed had been 20 miles per hour faster, the trip would have been made in 2 hours less. Find the speed.

Name ___________________________

Unit VIII – Quadratic Equations, Unit Test Form A
– Continued, Page 8 –

22. The length of a rectangle is twice the width. The area is 162 square yards. Find the length and width.

23. The standard form for a quadratic function is $ax^2 + bx + c = y$, given a solution (x, y). We want the data points $(1, 11)$, $(^-2, 2)$, and $(0, 0)$ to satisfy the function. Find values for a, b, and c to make a general quadratic function for which that is true.

Unit Test Form B

Name ______________________

Class __________ Date __________ Score ______

Unit VIII – Quadratic Equations

Find the solution set for each of the relations in problems 1-6.

1. $5n^2 - 8n + 3 = 0$

2. $x^2 - 49 = 0$

3. $p^2 - 11p = 0$

4. $4x + 20 = 0$

5. $7x^2 - 16 = 0$

6. $(v + 5)^2 = 13$

Name ______________________________

Unit VIII – Quadratic Equations, Unit Test Form B
– Continued, Page 2 –

Find the solution set for each of the relations in problems 7-10 by completing the square.

7. $x^2 + 6x - 16 = 0$

8. $m^2 + 4m - 7 = 0$

9. $2c^2 + 3c - 1 = 0$

10. $x^2 + 5x - 1 = 0$

Name ____________________________________

Unit VIII – Quadratic Equations, Unit Test Form B
– Continued, Page 3 –

Find the solution set for each of the relations in problems 11 and 12 by using the quadratic formula.

11. $x^2 - 4x + 13 = 0$

12. $3x^2 + 2x - 7 = 0$

In problems 13 and 14, find the value of the discriminant, and state the nature of the solutions.

13. $x^2 + 12x + 36 = 0$

14. $x^2 + 3x + 4 = 0$

Name ______________________________

Unit VIII – Quadratic Equations, Unit Test Form B
– Continued, Page 4 –

Show the solution set of each of the relations in problems 15-17 by finding the critical points on a number line and determining which interval(s) make the given relation true. Then write the solution set using set notation.

15. $z^2 - 4z - 12 > 0$

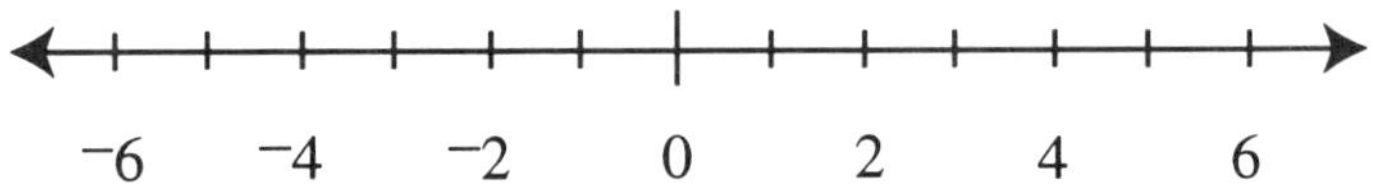

16. $x^2 - x - 2 \leq 0$

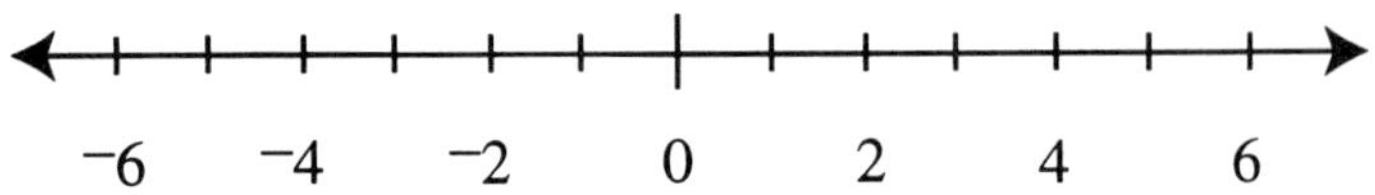

Name ________________________________

Unit VIII – Quadratic Equations, Unit Test Form B
– Continued, Page 5 –

17. $\frac{v-2}{v+5} < 0$

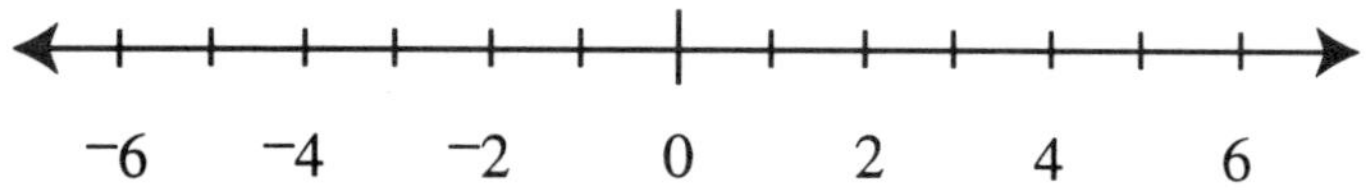

In problems 18 and 19, find the solution set of each relation by using a temporary substitution to "reduce" it to a quadratic equation in standard form.

18. $x - 10\sqrt{x} + 9 = 0$

19. $x^4 - 7x^2 + 12 = 0$

Name ________________________________

Unit VIII – Quadratic Equations, Unit Test Form B
– Continued, Page 6 –

20. Jane's age is 10 more than three times her son's age. In two years, her age will be 12 less than the square of her son's age. What is Jane's age now?

Name ___________________________________

Unit VIII – Quadratic Equations, Unit Test Form B
– Continued, Page 7 –

21. A boat travels 15 miles upstream and then turns around and travels 15 miles downstream. The total time for both trips is 4 hours. If the stream flows at 5 miles per hour, how fast does the boat travel in still water?

Name ___________________________

Unit VIII – Quadratic Equations, Unit Test Form B
– Continued, Page 8 –

22. The width of a rectangle is 5 inches less than its length. The area of the rectangle is 84 square inches. Find the dimensions of the rectangle.

23. The standard form for a quadratic function is $ax^2 + bx + c = y$, given a solution (x, y). We want the data points $(0, 0)$, $(4, 7)$, and $(^-4, 9)$ to satisfy the function. Find values for a, b, and c to make a general quadratic function for which that is true.

Cumulative Review Form A **Name** ______________________

Class ________ **Date** __________ **Score** ______

Algebra: A Complete Course
Units I – VIII

Simplify the following.

1. $\left(2+\sqrt{3}\right)\left(4-\sqrt{3}\right)$

2. $\sqrt{45}+8\cdot\sqrt{20}$

3. $\sqrt{24x^5}\cdot\sqrt{75}\cdot\sqrt{3x}$

4. i^{27}

5. $5i-(4-3i)$

6. $(4+i)(3-i)$

7. $\dfrac{10+i}{2+i}$

8. $(3+4x)(3+4x)$

Name ___________________________

Cumulative Review Form A
Algebra: A Complete Course, Units I – VIII

– Continued, Page 2 –

Use rational exponents to write a single radical expression.

9. $\sqrt[3]{x^2} \cdot \sqrt{x}$

Solve.

10. $3 = \sqrt{x-1}$

11. $\sqrt{2x+1} - 6 = 3$

12. $t^2 + 5t = 6$

13. $\dfrac{x+10}{2x} = \dfrac{3}{4}$

14. Determine whether $1+i$ is a solution to the equation $x^2 - 2x + 2 = 0$. Show your work.

15. Write 520,000 in scientific notation.

16. Factor completely: $48x^5 - 12x^3$

17. Find the union of $\{1, 7, 11, 13, 19\}$ and $\{1, 4, 13, 32\}$.

Name ______________________________

Cumulative Review Form A
Algebra: A Complete Course, Units I – VIII

– Continued, Page 3 –

Solve, and graph on a number line.

18. $7x - 2 < 16$ or $x + 3 > 5$

Solve.

19. $3x + 6 = 5 - (x + 8)$

20. $2t - 9 \leq 7$

Simplify. Write result with positive exponents.

21. $^{-}6x^{^{-}4}y^{0}$

22. $\left(\dfrac{12y^{5}}{8y^{^{-}4}}\right)^{^{-}3}$

Name ______________________________

Cumulative Review Form A
Algebra: A Complete Course, Units I – VIII

– Continued, Page 4 –

23. Chuck decided last year's garden was too big, so he decreased the length by two feet and decreased the width by three feet. What will be the total change in the area of his garden? (Hint: your answer will be an open phrase, not a numerical answer.)

24. From past experience, Jeff, who works for Kleen-Kars Incorporated, knows he can do a complete car detail in 4 hours. His assistant can do the same detail job in 6 hours. If they work together on the same car, how long should it take them to clean one car?

Cumulative Review Form B

Name ______________________

Class __________ Date __________ Score ______

Algebra: A Complete Course
Units I – VIII

Simplify the following.

1. $\left(6-\sqrt{18}\right)^2$

2. $\sqrt{108}+2\sqrt{3}+5\cdot\sqrt{12}$

3. $\sqrt{10x^3}\cdot\sqrt{21x}\cdot\sqrt{15x^4}$

4. i^{49}

5. $(8-i)-\left({}^{-}7-6i\right)$

6. $(2-3i)(5+4i)$

7. $\dfrac{3+4i}{3-i}$

8. $(5+6x)(5-6x)$

Name ______________________________

Cumulative Review Form B
Algebra: A Complete Course, Units I – VIII

– Continued, Page 2 –

Use rational exponents to write a single radical expression.

9. $\sqrt[6]{x} \cdot \sqrt[9]{x^5}$

Solve.

10. $7 = \sqrt{x-3}$

11. $\sqrt{4x-1} + 2 = 3$

12. $t^2 - 3t = 28$

13. $\frac{5}{6} = \frac{x-8}{4x}$

14. Determine whether $2+5i$ is a solution to the equation $x^2 + 29 = 4x$. Show your work.

15. Write 0.0042 in scientific notation.

16. Factor completely: $18x^7 - 50x^5$

17. Find the intersection of $\{2, 6, 10, 14\}$ and $\{1, 3, 4, 7, 8\}$.

Name ___________________________

Cumulative Review Form B
Algebra: A Complete Course, Units I – VIII

– Continued, Page 3 –

Solve, and graph on a number line.

18. $^{-}19 < 4x - 7 \le {}^{-}7$

Solve.

19. $0.2x - .38 = 1.3 - .6x$

20. $x = \frac{1}{3}y + 2$ for y

Simplify. Write result with positive exponents.

21. $8x^{0}y^{-1}$

22. $\left(\frac{6x^{-3}}{2x^{-2}}\right)^{-3}$

Name ________________________________

Cumulative Review Form B
Algebra: A Complete Course, Units I – VIII

– Continued, Page 4 –

23. Antonio designed his patio to be x feet long and y feet wide. Upon reflection, he decided it would be better to increase the length by five feet and the width by four feet. What is the amount of increase in the area? (Hint: your answer will be an open phrase, not a numerical answer.)

24. From past experience, a supervisor knows she can construct a certain porch in 9 hours. Her apprentice can construct the same porch in 12 hours. If the two work together, how long should it take them to construct the porch?

Name ______________________

Class ________ Date ________ Score ______

Comprehensive Assessment – Form A

Algebra: A Complete Course
Units I – VIII

1. Translate the following English sentence into a mathematical sentence.

"The result of seven times a number, decreased by 3, is the same as eighteen more than, four times the number."

2. Find the prime factored form of 90.

3. Find the least common multiple (LCM) of the set of numbers given below by factoring each number as a product of primes.

6 ______________________

12 ______________________ LCM ____________

8 ______________________

4. Simplify the following expression completely.

$$^{-}8 + 4x - 5$$

Name ____________________

Comprehensive Assessment Form A
Algebra: A Complete Course, Units I – VIII

– Continued, Page 2 –

Find the solution set for each relation in problems 5-7. Show each solution set on the accompanying number line, and then show it again, using set notation.

5. $2x-1<7$

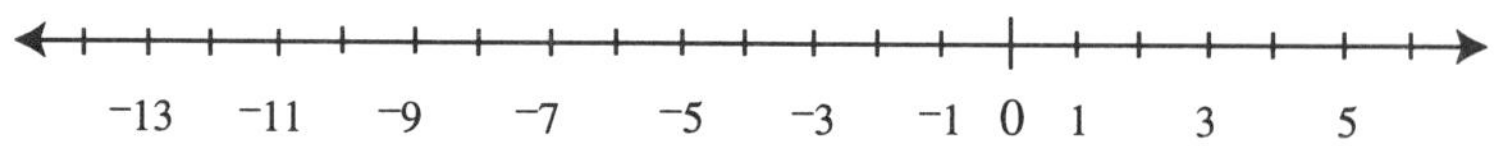

6. $n+3\geq 7$ or $2n-1\leq {}^{-}3$

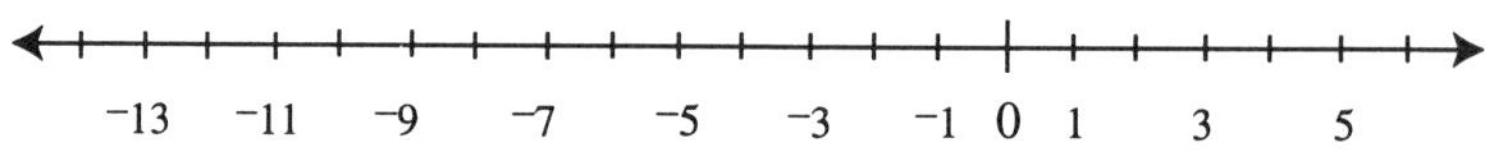

7. $|x+3|\leq 7$

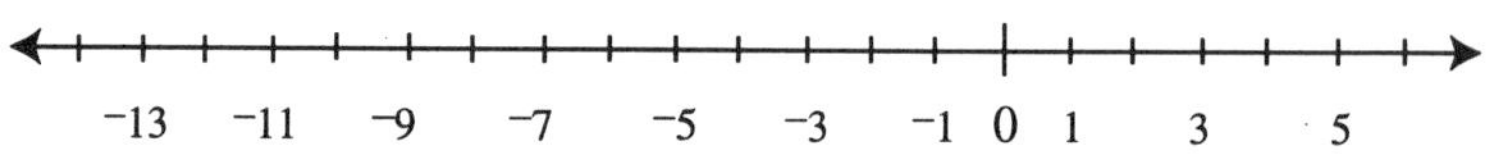

8. The sum of two consecutive odd integers is 184. Find the numbers.

Name ______________________________

Comprehensive Assessment Form A
Algebra: A Complete Course, Units I – VIII

– Continued, Page 3 –

9. Identify the quadrant (I, II, III, or IV) in which the point having coordinates $\left({}^{-}2,3\right)$ is located.

10. Solve the following relation for y.

$$5x + 2y = 20$$

11. Find the slope of the line through the points having the following coordinates:

(7, 5) and (9, 8)

12. Write the equation of the solution set line which has a slope of $\frac{2}{3}$ and a y-intercept of 4.

13. Graph the solution set line for the following equation.

$$x + 2y = 6$$

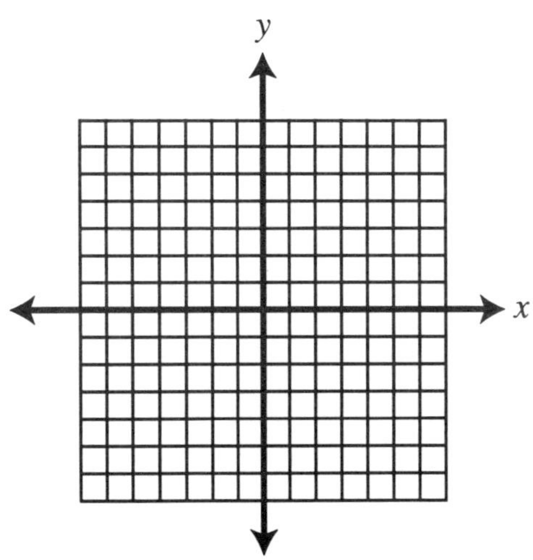

Name ________________________________

Comprehensive Assessment Form A
Algebra: A Complete Course, Units I – VIII

– Continued, Page 4 –

14. Find the solution set for the following system of equations, using the most appropriate method.

$$x + y = 1$$
$$2x + 5y = {}^{-}4$$

15. If you were making a graph of the solution set for the inequality $3x - 2y \leq 6$, would you shade above or below the line given by $3x - 2y = 6$? Explain why.

For the equation $x - 2y + 3z = 16$, complete each ordered triple in problems 16-18, so it will be a solution of the equation.

16. (______ , 5, 2)

17. (7, 0, ______)

18. (12, ______ , 4)

Name ______________________________

Comprehensive Assessment Form A
Algebra: A Complete Course, Units I – VIII

– Continued, Page 5 –

19. Evaluate completely: $3^2(2^3 - 2 \cdot 2^2) - (31 + 3^3)$

20. Simplify the following and write your answer in scientific notation.

$$\frac{2.7 \times 10^{-7}}{5.4 \times 10^3}$$

21. Find the following indicated sum:

$$(^-3x^2 + 4x - 9) + (x^2 - 2x + 3)$$

In problems 22 and 23, find the indicated products.

22. $5xy(4x^3y^2 - 5xy^3 + xy)$

23. $(x + 2)(2x^2 - 7x + 3)$

Name ____________________

Comprehensive Assessment Form A
Algebra: A Complete Course, Units I – VIII

– Continued, Page 6 –

Factor completely each of the expressions in problems 24-27.

24. $6m^2n + 8m^3n^2$

25. $x^2 - 8x + 7$

26. $25x^2 - 36$

27. $t^2 - 16t + 64$

Find the solution set for each of the polynomial equations in problems 28 and 29.

28. $(x-7)(2x+5) = 0$

29. $k^2 + 6k = 0$

30. State the restricted values for x in the following open phrase:

$$\frac{x+7}{(x+3)(1-x)} \cdot \frac{3x-8}{2x-10}$$

Name ________________________________

Comprehensive Assessment Form A
Algebra: A Complete Course, Units I – VIII

– Continued, Page 7 –

Perform the indicated operations in problems 31-34. Be sure to simplify your answer where possible.

31. $\dfrac{5n-5}{18n}\cdot\dfrac{9n^2}{n-1}$

32. $\dfrac{y^2-9}{4}\div\dfrac{y+3}{8}$

33. $\dfrac{6r-5}{8}+\dfrac{4r+2}{8}$

34. $\dfrac{2a+6}{a^2-9}+\dfrac{1}{a+3}$

Find the solution set for each equation in problems 35 and 36.

35. $\dfrac{2x}{5}+\dfrac{x}{10}=1$

36. $\dfrac{3}{5x}-\dfrac{2}{x}=\dfrac{1}{2}$

Name ______________________________

Comprehensive Assessment Form A
Algebra: A Complete Course, Units I – VIII

– Continued, Page 8 –

37. Solve the following inequality by considering the range of each factor relative to a number line. Be sure to note all restricted values, and show your solution set on the accompanying number line.

$$\frac{x}{x-4} \geq 0$$

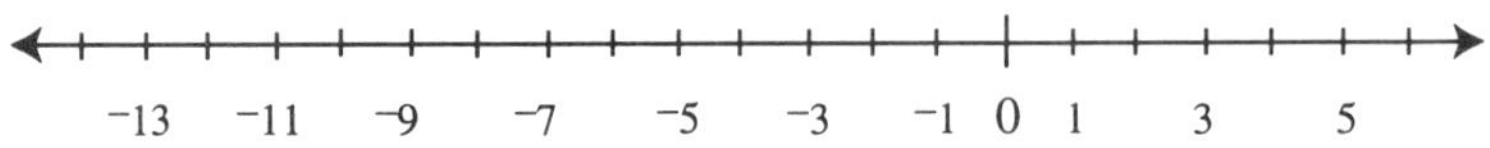

38. Solve the following relation for x.

$$W = \frac{x - B}{m}$$

Simplify completely each expression in problems 39-42.

39. $\sqrt{140}$

40. $\sqrt[3]{48x^3y^2}$

41. $5\sqrt{3} + \sqrt{12}$

42. $\sqrt{3}\left(3\sqrt{2} + \sqrt{3}\right)$

Name ________________________________

Comprehensive Assessment Form A
Algebra: A Complete Course, Units I – VIII

– Continued, Page 9 –

43. Find the solution set for the following relation:

$$\sqrt{5x-6}=3$$

44. Rationalize the denominator in the following fraction.

$$\frac{3+4i}{2-3i}$$

45. Solve the following quadratic relation by completing the square. Show your work.

$$x^2-6x-8=0$$

46. Use the quadratic formula to solve the following equation. Show your work.

$$3x^2-7x+2=0$$

Name ________________________________

Comprehensive Assessment Form A
Algebra: A Complete Course, Units I – VIII

– Continued, Page 10 –

47. Solve the following quadratic inequality by finding the critical points and determining which intervals satisfy the relation. Show your solution, first by a graph on the accompanying number line, and then by using set notation.

$x^2 - 3x - 10 < 0$

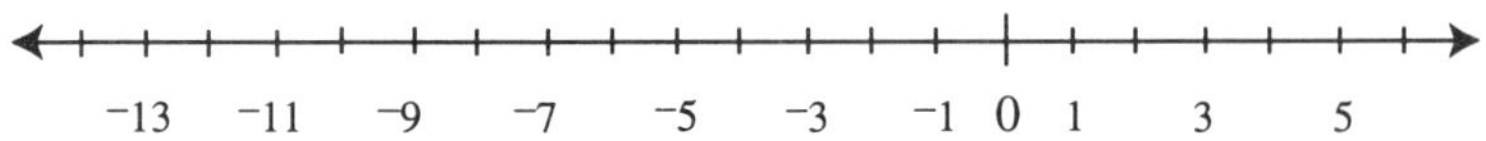

48. Solve the following equation by using a temporary substitution to "reduce" it to a quadratic equation in standard form.

$$x^4 - 9x^2 + 8 = 0$$

Name ________________________________

Comprehensive Assessment Form A
Algebra: A Complete Course, Units I – VIII

– Continued, Page 11 –

49. The hypotenuse of a right triangle is 26 feet long. The length of one leg is 14 feet more than the length of the other leg. Find the length of each of the legs.

50. A student opens a physics book to two facing pages. The product of the page numbers is 812. Find both of the page numbers.

Name ______________________________

Comprehensive Assessment Form A
Algebra: A Complete Course, Units I – VIII

– Continued, Page 12 –

51. Using the quadratic formula often produces solutions such as the following:

$$\frac{4 \pm 6\sqrt{3}}{4}$$

Many of these solutions can be simplified. Show that the following is an incorrect simplification by pointing out the error and then simplifying it correctly.

$$\frac{\cancel{4} \pm 6\sqrt{3}}{\cancel{4}} = \pm 6\sqrt{3}$$

52. The Pythagorean Relation is given as follows:

$$a^2 + b^2 = c^2$$

Explain this relation using the terms "leg" and "hypotenuse" as the following diagram illustrates. Then give a numerical example of the relation.

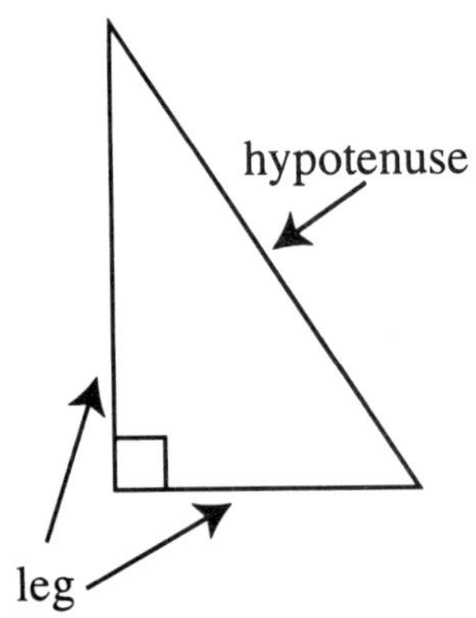

Name ______________________________

Class __________ Date __________ Score ______

Comprehensive Assessment – Form B

Algebra: A Complete Course
Units I – VIII

1. Find the greatest common factor (GCF) of the set of numbers given below by factoring each number as a product of primes.

   36 ______________________

   48 ______________________ GCF ____________

   60 ______________________

2. Evaluate the open phrase $2xy - 5y$ when $x = 3$ and $y = {}^-2$.

3. Evaluate $|K| + |M|$ when $K = 3$ and $M = {}^-6$.

4. Graph the given set on the number line provided.

   {integers greater than 2}

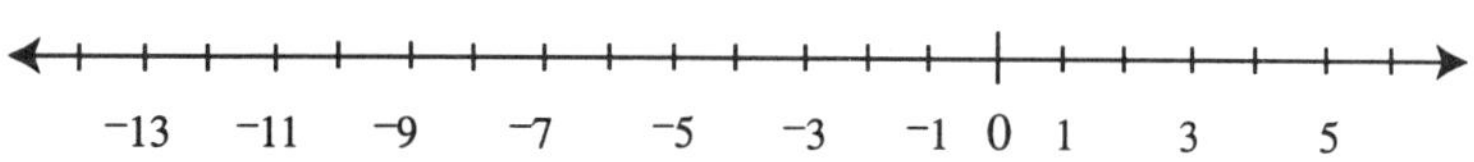

Name ______________________________

Comprehensive Assessment Form B
Algebra: A Complete Course, Units I – VIII

– Continued, Page 2 –

5. Solve the following open sentence. Show your solution on the number line provided and in set notation.

$7(2a-4)=2(a+4)$

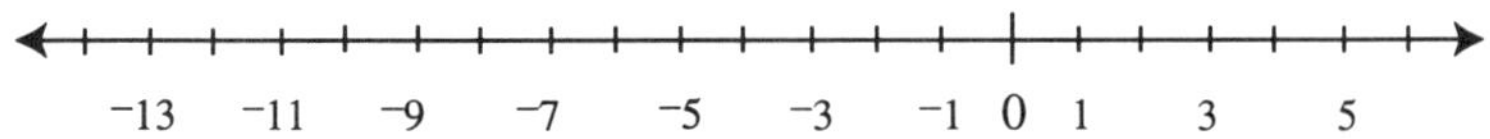

6. Find the solution set for the following compound sentence. Show your solution set on the accompanying number line and in set notation.

$3m-4\leq 7m+12$ and $8m-7\leq 25$

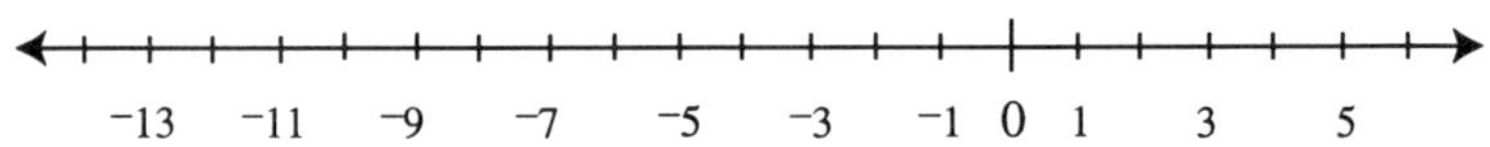

7. Find the solution set for the following absolute value relation. Show your solution set on the accompanying number line and in set notation.

$|4p-3|=7$

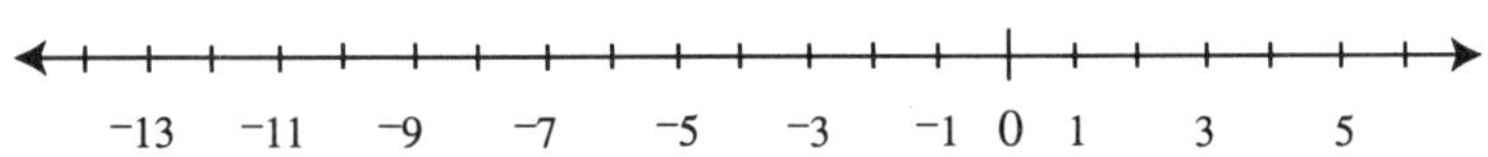

8. Use one of the three basic relation symbols $(=, >, <)$ in the following to make a true statement of relationship.

$$\frac{3}{4} \; \_\_\_\_ \; \frac{2}{3}$$

Name ______________________________

Comprehensive Assessment Form B
Algebra: A Complete Course, Units I – VIII

– Continued, Page 3 –

9. For the given function and domain, find the range.

$f(\ \ ) = 4 - 3(\ \ )$

The domain is {0, 1, 2}

10. Determine the slope for the solution set line $3x + y = 5$.

11. Determine the x-intercept for the solution set line $5x - 6y = 60$.

12. State the slope of the line perpendicular to the line whose slope is $\frac{4}{3}$.

Name ______________________________

Comprehensive Assessment Form B
Algebra: A Complete Course, Units I – VIII

– Continued, Page 4 –

13. Graph $y = 2|x - 2|$

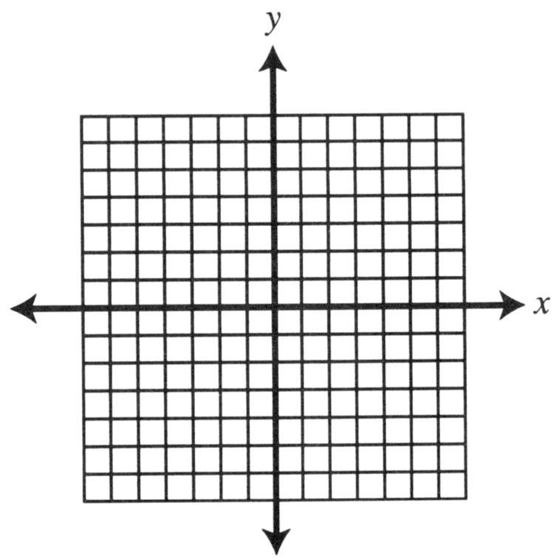

14. Solve the following system of simultaneous linear equations.

$$3x - 4y = 5$$
$$^{-}3x + 6y = 7$$

15. If you were making a graph of the inequality $y > {}^{-}2x + 5$, would the boundary line be solid or broken?

Name ______________________________

Comprehensive Assessment Form B
Algebra: A Complete Course, Units I – VIII

– Continued, Page 5 –

16. Graph the given system of inequalities.

$$y \le 2x + 3$$
$$x > {}^{-}1$$

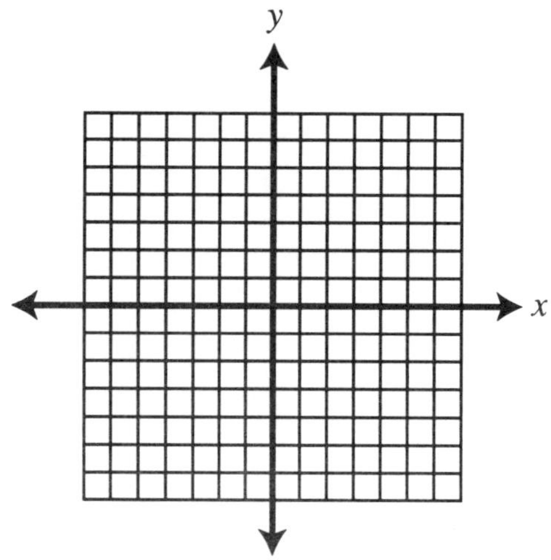

17. For the following equation, determine whether the given ordered triple is a solution.

$$x - 3y + z = {}^{-}1 \quad \left({}^{-}2, 3, 10\right)$$

18. Evaluate: $(9 - 3 \times 2)^4$

19. Simplify: $\dfrac{48a^3b^7}{12a^4b}$

(Write your answer with positive exponents.)

20. Simplify the following open phrase.

$$3 + \left(2s^2 - 5t - 1\right) - 2\left(s^2 + 2t - 4\right)$$

Name ______________________________

Comprehensive Assessment Form B
Algebra: A Complete Course, Units I – VIII

– Continued, Page 6 –

For each of the following, find the indicated product.

21. $(x+4)(2x+6)(x+4)$

22. $(7x)(3x^2+5x-1)$

Factor each of the following completely.

23. $9x^2-64$

24. x^2-9x+8

25. $y^2+24y+144$

26. x^3-16x

Find the solution set for the following polynomial equations.

27. $2x^2-15x+7=0$

28. $(x-8)(3x+1)=0$

Name ____________________________________

Comprehensive Assessment Form B
Algebra: A Complete Course, Units I – VIII

– Continued, Page 7 –

29. $x^2 - 6x = 0$

30. Solve the following inequality by considering the range of each factor relative to a number line.

$x^2 + x - 6 \geq 0$

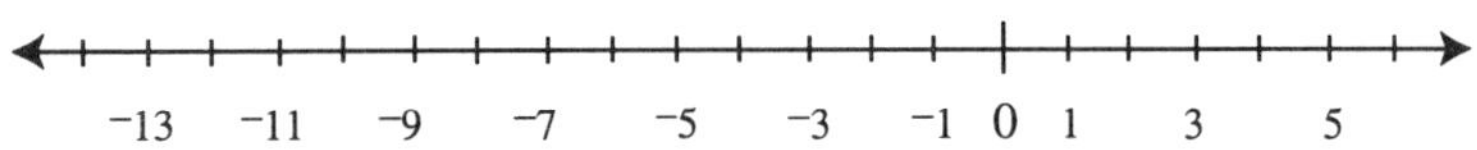

Perform the following operations in each expression.

31. $\dfrac{y^2 - 64}{y^2 - 8y + 12} \div \dfrac{4y - 32}{y^2 - 4y - 12}$

32. $\dfrac{3 + \frac{1}{y}}{3 - \frac{1}{y}}$

Name ________________________________

Comprehensive Assessment Form B
Algebra: A Complete Course, Units I – VIII

– Continued, Page 8 –

Solve.

33. $\dfrac{2}{m-1}=\dfrac{3}{m+1}$

34. $\dfrac{6}{a+3}+\dfrac{4}{a-3}=\dfrac{14}{a^2-9}$

35. $A=\dfrac{KB}{C}$ for K

Simplify.

36. $\sqrt{40}+\sqrt{90}$

37. $\sqrt[3]{9x^5y^6}\cdot\sqrt[3]{3x^7y^5}$

38. $2\sqrt{80}$

39. $x^{\frac{2}{3}}\cdot x^{\frac{1}{2}}$

Name ____________________________________

Comprehensive Assessment Form B
Algebra: A Complete Course, Units I – VIII

– Continued, Page 9 –

40. $3\sqrt{x^3} + 2x\sqrt{x}$

41. Solve for x: $\sqrt{4x+1} = \sqrt{7x-8}$

42. Rationalize the denominator in $\dfrac{6+\sqrt{2}}{3-\sqrt{2}}$.

43. Solve the following quadratic relation by completing the square.

$$x^2 + 4x - 5 = 0$$

Name ________________________________

Comprehensive Assessment Form B
Algebra: A Complete Course, Units I – VIII

– Continued, Page 10 –

44. Use the quadratic formula to solve $^{-}2x^2 + 6x + 3 = 0$.

45. Solve the following quadratic inequality by finding the critical points and determining which intervals satisfy the inequality. Show your solution, first by a graph on a number line, then by using set notation.

$x^2 + 2x - 3 < 0$

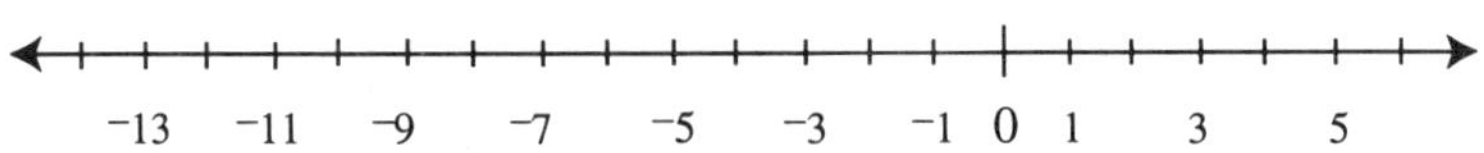

Name ______________________________

Comprehensive Assessment Form B
Algebra: A Complete Course, Units I – VIII

– Continued, Page 11 –

46. Solve the following equation by using a temporary substitution to "reduce" it to a quadratic equation in standard form.

$$x - 3x^{\frac{1}{2}} - 4 = 0$$

47. The width of a rectangle is 4 feet less than the length. The area is 10 square feet. Find the length and the width.

Name ______________________________

Comprehensive Assessment Form B
Algebra: A Complete Course, Units I – VIII

– Continued, Page 12 –

48. A student opens a mathematics book to two facing pages. The product of the page numbers is 2,162. What are the two page numbers?

49. An expression such as $\frac{3x(x-7)}{6x(x-7)(x+2)}$ was simplified on a recent test to $2(x+2)$. Was this correct? Why or why not?

50. The area of a triangle is given by the expression $A = \frac{1}{2} \cdot bh$. Write this equation in words.

